ELLEN B. LAGERWERFF &
KAREN A. PERLROTH

mensendieck your posture

encountering gravity the correct and beautiful way

LINE DRAWINGS BY ELLEN B. LAGERWERFF
ANATOMY DRAWINGS BY KAREN A. PERLROTH

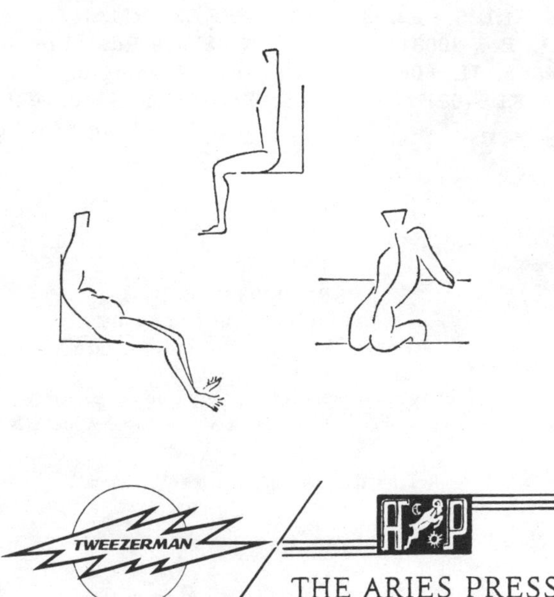

Acknowledgments

We express our sincere gratitude
to John M. Lagerwerff, M.D., for
his unfailing co-operation

To Molly Lewin, for her valuable
suggestions and practical assistance
in editing our book.

Revised Edition 1982

CO-PUBLISHED BY:

THE ARIES PRESS
P. O. Box 30081
Chicago, IL 60630
(312) KI 5-0717

TWEEZERMAN
W. Shore Rd., Box 761
Port Washington, NY 11050
Toll Free: 1-800-645-3340
In N. Y.: (516) 944-9726

ISBN 0-933646-19-4
previously published by
Anchor Press/Doubleday under
ISBN: 0-385-02112-7
Library of Congress Catalog Card Number 72-97093

Printed in the United States of America

FOREWORD

My training is as a psychiatrist and I have no special expertness in kinesiology. I have, however, had significant personal experience as a beneficiary of the Mensendieck method, and that is my reason for writing this foreword.

I grew up with an inadequate posture, as all too many people do. The consequences were unfortunate. Collapsed disks in neck and back placed pressure on sensory nerves which caused painful and fatiguing muscular spasms. The corrective Mensendieck Routines, prescribed by Mrs. Ellen B. Lagerwerff, one of the authors of this book, helped relax the muscular spasms. Despite the presence of the spinal deterioration, I was able to improve my posture so appreciably that the faulty pressure on the sensory nerves was eliminated. I could sleep comfortably and move with a body flexibility that I was afraid I had lost. A number of my patients and friends have had comparable experiences working with the authors of this book.

Mrs. Lagerwerff received her higher education in the Netherlands and became qualified as a Mensendieck specialist after graduating from the rigorous training program of the Mensendieck Institute in Amsterdam, Holland. She practiced in The Hague for nineteen years before emigrating to the United States in 1959 with her physician husband and family. Her daughter, Mrs. Karen A. Perlroth, who is the coauthor of this volume, graduated from the same institute in 1966. Her husband is also a physician. Mrs. Lagerwerff and Mrs. Perlroth are in private practice in California, Palo Alto and Portola Valley, respectively.

Mensendieck takes a unique approach. While the requirement for conditioning the body is recognized in common with the recommendations of others, the specific techniques and the specific aims place this method in a class by itself.

Although bodily conditioning by this method will improve athletic skill, there is no emphasis upon extraordinary muscular performance, valuable as this might be for some people in some contexts. The emphasis is on correct and graceful body movements throughout all everyday activities.

How many of us are aware that our daily life activities can be performed in a way beneficial to the body? A glance at the table of contents of this book can serve as a reminder of how one uses the body in a round of daily tasks. Even the activities carried on within a physically active life often tend toward a faulty postural alignment. For example, much that we do is with the arms in front of the body, bringing the shoulders forward while moderately collapsing the chest. Stooped shoulders, a relaxed abdomen on which fat settles, and eventually backaches may be a consequence. If one is to keep one's body well aligned and in excellent condition, one needs to stand, sit, and move properly; in this way little-used muscles will be given an opportunity to function. This book is geared to facts such as these.

The reader will have his eyes opened to numerous unthought-of body-conditioning essentials, frequently of the kind that will make him ask: "Why didn't I think of that myself?" But then, as he moves on into the more advanced techniques of the later chapters, he will recognize that everything is based on detailed knowledge of anatomy and of muscular insertions, brought to practical usage through a high level of ingenuity and wisdom.

The exposition of many basic principles of Dr. Mensendieck's system is fresh and enriched by the authors' broad experience over many years. Their primary aim is to help each reader apply these principles, so that, as he carries out his daily tasks, the human machine at his disposal will function smoothly and without pain.

FOREWORD

As a psychiatrist I am of course aware of how much a good body image contributes to mental health. The series of routines described in this book can help one attain such an image.

A rich reward in health, comfort, and *joie de vivre* lies ahead for the reader who will work his way through the book, learn from it, and practice what it preaches.

Josephine R. Hilgard, M.D., Ph.D.
Clinical Professor of Psychiatry,
Stanford University

"Within the framework of its skeletal structure,
you can sculpture the shape of your body at will."

Bess M. Mensendieck, M.D.

CONTENTS

INTRODUCTION

The purpose of this book is to acquaint you with certain basic principles which, when understood and applied, can mold your body into a trim and well-balanced structure.

Posture is usually defined as the relative position and alignment of the various masses of the human body. How you actually superpose your body masses is entirely up to you.

The human body, like an architectural structure, is subject to the laws of gravity. The big difference lies in the engineering requirements. A building's separate masses have to be correctly arranged only once. Man, however, has to constantly "engineer" in order to maintain the correct relationship between the various masses of his mobile structure. These masses tend to move continually in relation to one another when you are standing still. This tendency increases markedly the moment you relocate the entire body or one of its masses.

Each human structure will start to sag and to bulge as soon as you, the "resident engineer," allow its mutually movable parts to yield to gravity. Nobody is born with such deformities as flatfeet, bowlegs, a sway-back, a potbelly, stooped shoulders, a hunchedback, or a dowager's hump. Such deformities result from the engineer's long-time neglect of his obligations to keep up his physical structure.

In the course of our evolution the function of the muscles in sustaining posture has become increasingly important. Four-legged animals hardly ever develop deformities as the result of faulty posture. If the human trunk were also carried horizontally on four legs, we would have to be much less concerned about posture. The quadrangle between the four feet would give us a much larger and firmer foundation than the small square outlined by our two feet. Since the body structure would not be so tall, the downward compression of the various body masses would be far less, and their total weight would be carried by four supports instead of two.

The smaller a structure's foundation, the more likely it is to fall over. The taller a structure, the more effort is required to hold it upright and the more pressure will be exerted by the higher masses upon the lower masses and on the foundation. Hence, each living creature born as man has to be concerned about how to poise and how to sustain his tall and flexible stature on its small base.

Man has to learn how to correctly distribute the weight of the upper structures over the lower structures of his body and how to simultaneously distribute his total weight over his feet. Once he has mastered proper carriage while standing, he can become skillful in moving functionally. This requires constant adjustments in muscular activity to properly counter the effects of gravity. Normally, man comes into the world with a potentially perfect neuromuscular system, which is the only equipment required to cope with his body's mechanics. All you have to do is learn to correctly select specific muscles for specific tasks. This gives your musculature a fair chance for balanced growth.

Contrary to popular belief, a healthy muscle does not necessarily have to be overly strong and bulky. It must be able to alternate smoothly between adequate tension and relaxation, and it must possess resiliency or the ability to return to its former shape after distortion. In a balanced musculature, opposing muscle groups have approximately the same resilience.

Fortunately, it is never too late to start balancing your muscles through proper training. Neglected muscles must be trained to gain tone, whereas the tone in muscles which are excessively used must be decreased. Thanks to the genius of Bess M. Mensendieck, M.D., nobody has to figure out for himself how to effectively re-educate his musculature.

Dr. Mensendieck was a sculptor who became a physician in order to find out exactly what man would have to do to sculpture his own body into a structure he could be proud of and that would serve him well. During many years of research in kinesiology (the science of human posture and motion), she analyzed in minute detail the individual functions of every bone, joint, and muscle, as well as their intricate interactions. Dr. Mensendieck arrived at the following conclusions:

1. No daily half hour or any periodic physical exercise can ever create a balanced musculature if the muscles are not properly used the rest of the time.
2. Correct posture and freedom from pain due to faulty posture can only be achieved when all muscles are used in accordance with their anatomical functions and the laws of body mechanics.

Thus, Dr. Mensendieck became convinced that the primary exercise anyone needs is the maintenance of correct posture during normal daily activities. She designed numerous ingenious techniques and developed a specific method of training through which everyone could become aware of how to construct and maintain the correct alignment of his body masses.

In the course of their over thirty years of experience as Mensendieck kinesio-specialists, the authors have come to the realization that there are a number of problems related to posture which almost all people have in common. In this book, we have restricted ourselves to the discussion of the prevention and the correction of these common causes for faulty posture and related pains.

Whenever you have a persistent pain, you should, of course, see your physician for a thorough physical examination and possible treatment. If he diagnoses your pain as being caused by poor posture, which thus resulted in an imbalanced musculature, the Mensendieck Techniques described in easy-to-follow steps in this book can help you.

Ideally, you should practice the Mensendieck Routines using three mirrors, one in front of you, one on the side, and one obliquely placed in the rear. If three mirrors are not possible, try to have one in the front and one in the rear. You must be able to catch your rear image in the front mirror, which should preferably be a large wall mirror.

It is essential to practice basic Mensendieck with the body undressed, or scantily dressed in bikini panties or tight-fitting swimming trunks. You must be able to observe the individual muscles in action so that you can be sure that you are selecting the right ones to hold you trim. You should not wear a leotard because such a garment lulls you into a false sensation of being held snug although your muscles may be lazy.

IT IS IMPORTANT THAT YOU WORK YOUR WAY THROUGH THE CHAPTERS IN SEQUENTIAL ORDER.

Read each chapter very thoroughly and let your mind absorb what is being explained in the text and in the diagrams and anatomical drawings. Then position yourself between the mirrors (or if only one is available, face it) in the way you habitually stand. Subsequently "build" your posture in exact accordance with the Mensendieck Technique just studied. Do this time and again, until you are completely familiar with the movement.

Make it a point, from now on, always to apply the basic principles of the learned Techniques in subsequent Routines as well as in your everyday activities. In time you will do so habitually, which means that you will then be "exercising" your body continually in the most constructive manner.

Ellen B. Lagerwerff Karen A. Perlroth

Members of the Nederlandse Mensendieck Bond,
California Branch USA and of the
Academy of Mensendieck Specialists

A FEW WORDS CONCERNING PROFESSIONAL INSTRUCTION

This book has been explicitly written as a self-instruction guide for beginners. Complete mastery of the contents of this book, however, will not impart sufficient knowledge to anyone, including practitioners of other allied disciplines such as physical therapists, for teaching or instructing pupils, students, or patients in the preventive, corrective, or rehabilitative Mensendieck specialty. A qualified instructor must have far broader and deeper knowledge of Mensendieck. For example, at the time of this writing the Dutch Mensendieck Teachers Association requires that enrollees at the Mensendieck Institute have a Bachelor of Science degree or its European equivalent, and that each student must successfully complete the three-year curriculum in order to become a certified Mensendieck instructor. In the Netherlands, the Nederlandse Mensendieck Bond was legally authorized by Bess M. Mensendieck personally to validate Mensendieck Teachers' Certificates or Diplomas on her behalf. No Mensendieck Teachers Association was established by Dr. Mensendieck in the United States for reasons set forth in the chapter entitled "The Mensendieck System—Background Information." In the opinion of the authors, no courses comparable to those offered by the European Mensendieck Institutes are given in the United States at the present time. Therefore, the authors recommend the services of Mensendieck specialists certified by one of the European Institutes. A list of the limited number of these specialists practicing in the United States may be obtained by self-addressed envelope from the Academy of Mensendieck Specialists, P. O. Box 9450, Stanford, California **94305-0270**

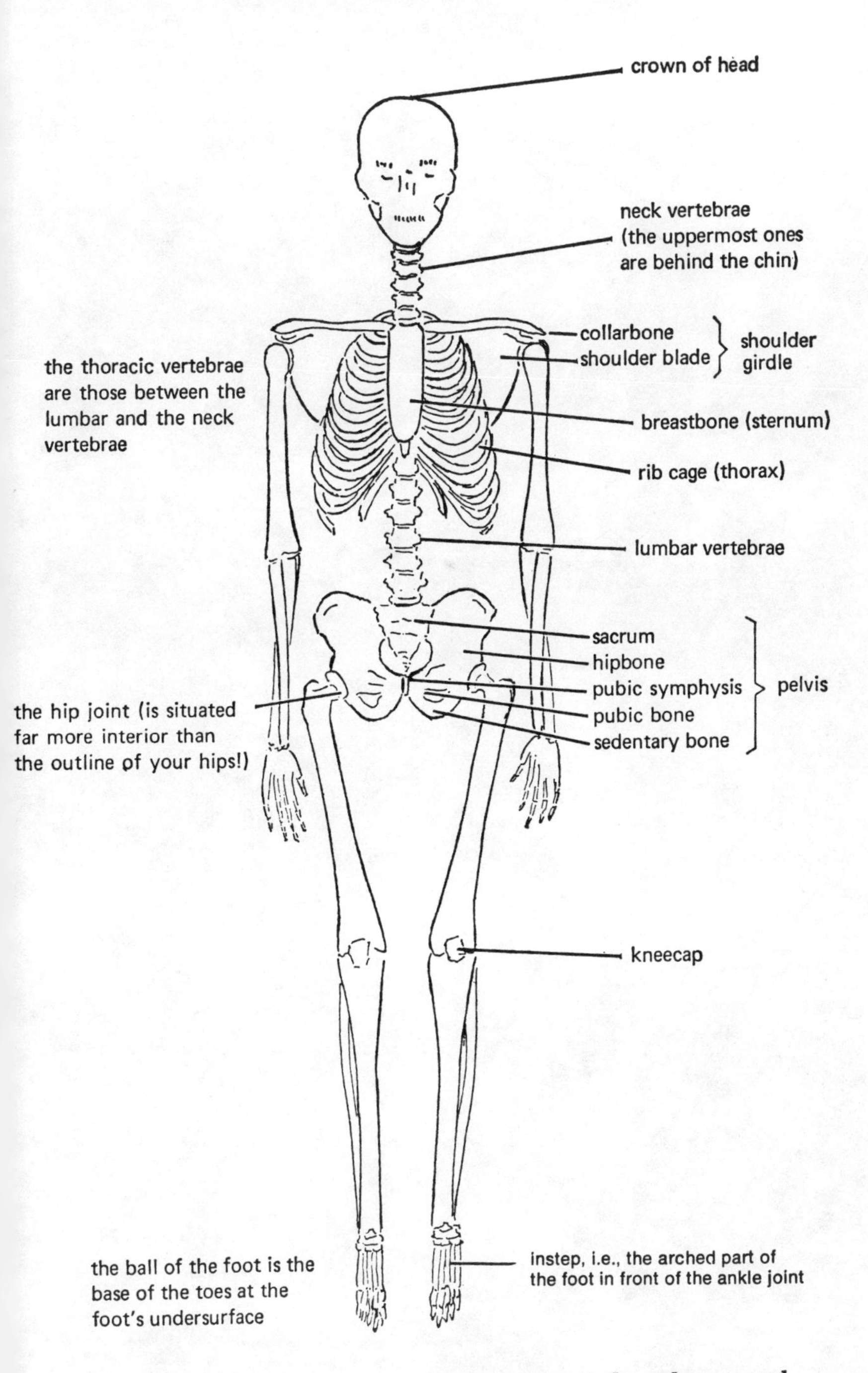

The human skeleton as seen from the front; its parts indicated as essential for this book.

IMPROVING YOUR POSTURE AND BASIC MOVEMENTS

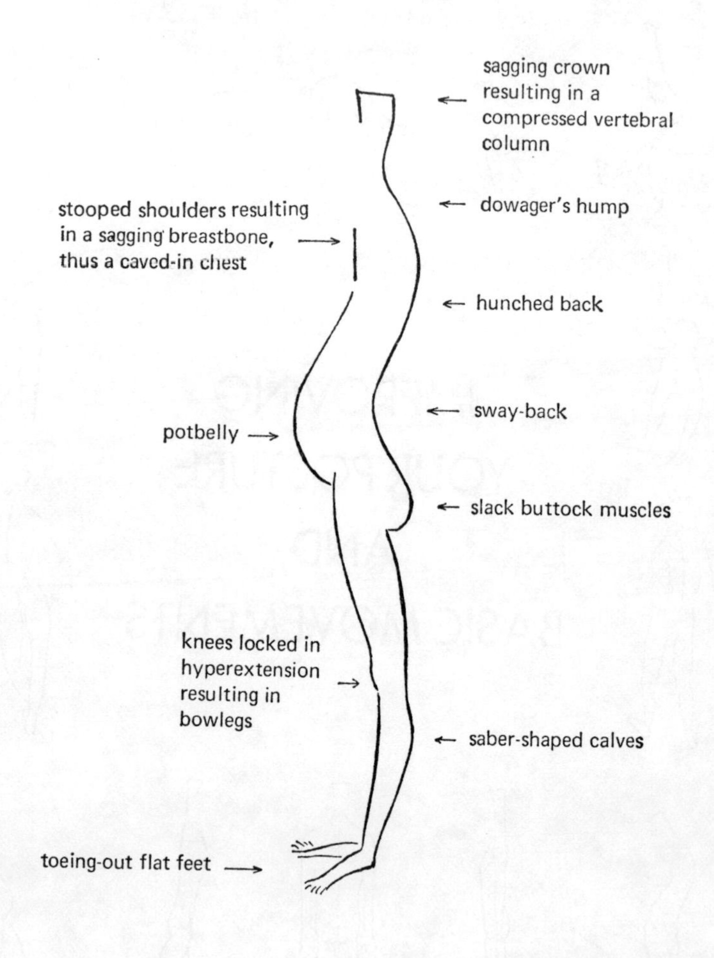

Diagram of the most common defects in posture

I. STANDING AND WALKING

In order to understand the reasons for the Techniques which lead to proper standing and walking, it might be helpful to keep in mind some of the most common defects in carriage. The ones discussed here are: outward-pointing toes, locked knees, slack buttock muscles, improperly contracted abdominal wall, and faulty positioning of the shoulder blades and breastbone.

The habit of "toeing-out" makes it all but impossible for a person to position his various body masses properly. When your toes point outward, your knees do the same. Actually, both of your legs are then rotated outward (rolled out) in the hip joints. In this position the distance between the beginning and the end of your buttock muscles is shorter than normal. This causes these muscles to hang slack, which makes it very difficult to contract them continually while standing and walking. Effective use of the buttock muscles is of paramount importance for the proper positioning of many of your bones, even those in your feet.

Fatty tissue tends to settle on slack, thus lazy, muscles. Start to reverse this trend right now by always placing your feet parallel. This will be further explained under Mensendieck Technique 1, the Well-balanced Stance.

Knee joints which are locked too tightly are easily vulnerable to harm. Such a stance also causes saber-shaped calves and bowlegs. And, consequently, when you stand with the bones in your legs improperly aligned, you are forced to distort the alignment of upper-body masses in order to maintain balance.

Never straighten your knees all the way; instead, bend them ever so slightly.

We have found that whenever a person is asked to contract his abdominal muscles, he tries to accomplish this by "sucking in" the abdominal wall while inhaling and raising the chest. This stretches most of these muscles rather than contracting them.

How to properly contract abdominal muscles is explained in Technique 1; why exhaling should accompany firm abdominal contractions is clarified in Technique 4.

Man tends to allow his shoulder girdle to suspend itself from the only point at which it is joined to the rest of his skeleton, at the top of the breastbone (sternum). The positioning of your breastbone in relation to the pelvis (for effective functioning of your abdominal muscles) and in relation to the spine (for efficient breathing) is essential for proper carriage. Nobody can tighten his abdominal muscles and breathe adequately with a caved-in-chest.

The correct alignment of your sternum is dependent upon proper carriage of the shoulder blades and the ideal alignment of the vertebral column, as explained in the Well-balanced Stance, which is the most fundamental Mensendieck Technique.

Mensendieck Technique 1: THE WELL-BALANCED STANCE

• Position your feet parallel and exactly below your hip *joints* by placing the balls of your feet almost two inches apart, toes pointing straight ahead, and by placing your heels about three inches apart so that they point straight backward. (The heels are now obscured from view in your front mirror image.)

• Raise your pelvis in its hip joints by tucking your buttocks firmly together and slightly under.

Sway-backs should firmly emphasize the downward pull (as if they are drawing an imaginary tail between the legs) in order to correct this defect and move the pelvis to its upright position. Those with too-straight backs should not pull the buttocks under at all, but just together. In all cases the lower back (lumbar curvature) must now be shallowly concave or "hollow."

• Contract your abdominal muscles firmly upward, while maintaining a tall stance.

The tightening of your buttocks brought about a forward-and-upward movement of your pubic bones. "Hooking up" these bones to your sternum is the task of your belly muscles. (How to bring about the proper positioning of your sternum, so that your belly muscles can contract effectively, will be explained shortly.) Your buttocks and abdominal muscles have to learn to work in perfect co-ordination to

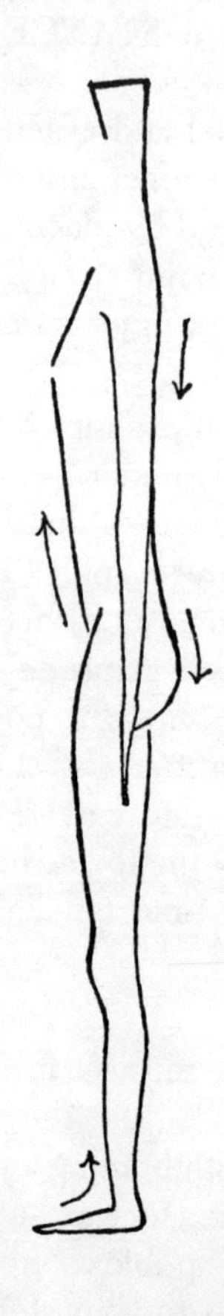

Diagram of the body properly carried in the Well-balanced Stance resulting in the proper positioning of your arms, i.e., thumb on trouser seam. NOTE: The arrows indicate the correct engagement of principal muscle groups.

counteract gravity and carry your pelvis in ideal position in its hip joints. When this is achieved, the contents of your abdominal cavity will no longer rest against its protruded and weakened wall, but rather on the tough pelvic floor.

• If during the above two phases your feet remained parallel and your knee joints were not too tightly locked, your legs generally and knees specifically will now automatically have assumed their correct relationship with your feet and your pelvis.

If you have been in the habit of standing flat-footed and/or bowlegged, these postural faults are now eliminated. Make sure, however, that your heels and the bases of your big toes remain in firm contact with the floor; loading your outer foot margins is incorrect.

• The sternum can now be brought in correct position, diagonally forward-up, by drawing your shoulder blades toward the vertebral column and downward, and by allowing them to "squeeze" the lower thoracic vertebrae (in the middle of your back) forward.

Be sure to pull with the muscles in the middle of your back, rather than to push with your arms.

• Your vertebral column should now have assumed correct alignment, i.e., shallowly concave in its lumbar section, slightly convex in its thoracic part, and the neck shallowly concave.

Succeeding in this depends upon the present potential for flexibility in your spinal column. If you have allowed your spine to become more or less rigid, do the best you can and work for gradual improvement. Perseverance is definitely called for in these cases. As long as your spine allows some movement, you can eventually regain its flexibility.

• Extend your neck upward to your crown.

The secret for the prevention of a double chin and/or a dowager's hump is to never position your face or your chin, but rather your neck. Your crown, not your forehead, should be your topmost spot.

• Your head should now "rest," perfectly poised, on your spine.

If you succeeded in aligning your body masses in perfect balance, from the feet up, hardly any muscle action will be required to carry your head.

• Your arms should now automatically have come precisely alongside your body, as a result of the proper positioning of all the other body masses.

The elbows should be straight and the palms of your hands should face backward.

• The weight of your well-aligned body must be carried equally distributed over your heels and the bases of your big toes.

The resulting, slightly diagonal stance is easy on your feet since each of the four points of support carries one-fourth of your body weight. This stance also stimulates all your posture-sustaining muscles into normal activity.

Disciplining yourself habitually to stand well balanced in your daily life will greatly enhance your appearance and well-being. Construct your posture from the feet up, slowly in the beginning. Once mastered, you will be able to do it in a split second, and many of the other Mensendieck Routines will be relatively easy.

Mensendieck Technique 2: THE ONE-LEGGED STANCE

People frequently stand on one leg instead of on both, with the pelvis in a sideways tilted position, thereby suspending most of the body weight from the hip joint of the leg on which they stand. This throws the body completely out of normal alignment and its musculature out of balance. To make matters worse, the prevailing tendency is to always suspend from the same leg. If people alternated regularly between their two legs, the harm to the body would be less.

In order to stand correctly on one leg, you must maintain tension in the muscles of the hip of your standing leg:

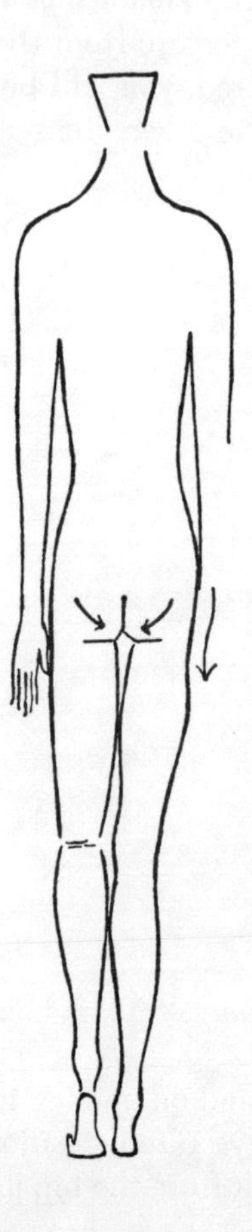

In order to stand correctly on one leg, you must maintain tension in the muscles of the hip of your standing leg as well as in your buttocks and belly muscles.

PREPARATORY POSITION: The Well-balanced Stance

• Carefully transfer your weight to one foot by mainly moving your upright pelvis sideways.

While during the Well-balanced Stance, the imaginary vertical line through your nose and jugular notch and pubic symphysis touched the floor between your two feet, this vertical line will now touch the one loaded foot. In other words, your upper torso should not curve to one side.

• With the feet in parallel position, place the unloaded one on its toes next to the heel of your standing leg.

The pelvis remains upright and frontal; do not budge in the hip joint of your standing leg.

• Maintain this One-legged Stance for a while, to gain distinct awareness of your "corset muscles," i.e., those in abdomen, buttocks, and hips.

• Place the unloaded foot back next to the loaded one, parallel, with about two inches between the bases of your big toes.

• Carefully distribute your weight equally over both feet.

Repeat these five phases, alternating the standing leg.

By executing the One-legged Stance off and on throughout your daily activities, the muscles along the inside and outside of your thighs will remain trim. Our legs hardly ever move sideways; frequent standing correctly on one leg is ideal to counteract that. Remember to demand balanced activity from your musculature by regularly alternating between your left and your right leg.

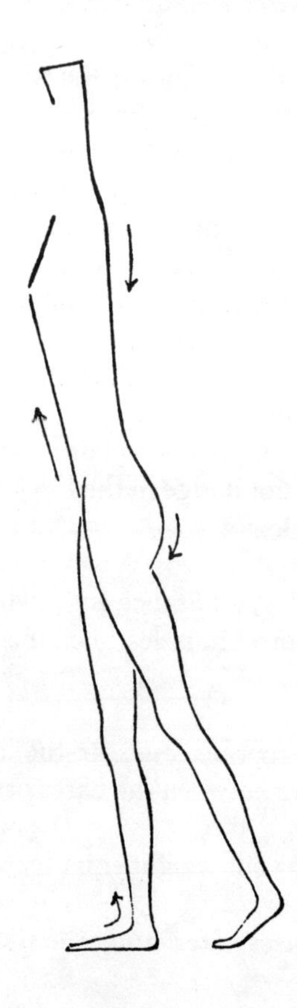

. . . follow your sternum diagonally forward-up; keep your weight ahead of you by propelling yourself forward with your pinching buttocks. The diagram shows the very moment in which your rear knee will have to subsequently bend in order to take the next step. . . .

Mensendieck Technique 3: WALKING

Subconsciously, man seems to interpret the verb "to walk" primarily as an activity of the legs, the taking of steps in order to displace the body's mass in the forward direction. Consequently, this activity is initiated by sticking out one leg, while leaving the body behind. The latter then has to be pulled forward until its weight is on the forward foot. Only then is it possible to stick out the other leg for the next step, and so on. Walking in this manner requires a tremendous amount of energy and is time-consuming; the body has to be dragged along almost entirely by sheer muscle effort. Even if the body was properly aligned before the walk was initiated, it is all but impossible to maintain correct alignment the moment one starts walking in this inefficient manner.

Rather than allow your legs to drag your body along, it is your body that should carry your legs with it:

PREPARATORY POSITION: The Well-balanced Stance.

• Keeping the body tall, thus well aligned, carefully shift its weight forward in the ankles by pinching your buttocks together and under.

• Due to the displacement of the center of gravity forward, your weight will "invite" you to take the first step; your heels will almost spontaneously leave the floor. Just prior to this very moment, your first step must be initiated before any raising in the ankles can take place, which would result in approaching the ceiling with your head.

• Subsequent steps will follow automatically if you stay well aligned and always keep your weight ahead of your forward foot.

Just respond to what wants to happen. Be your tallest, follow your sternum diagonally forward-up, keeping the crown of your head equal distance from ceiling. Continue to keep your weight ahead of you by propelling yourself forward with your pinching buttocks.

You will now experience yourself walking with ease, light-footed and fast; your gait should now feel as if you are floating.

II. BREATHING

Efficient breathing is vital to maintain optimum metabolism in all parts of the body, particularly in the muscles and the brain. It is distressing that most people breathe very poorly; their rib cage scarcely expands because they inhale mainly with the auxiliary breathing muscles instead of the principal ones.

It is our experience that men usually inhale by contracting the diaphragm to expand the lungs, thereby forcing the contents of the abdominal cavity forward. This manner of breathing ventilates the bottom of the lungs primarily, and will sooner or later cause a potbelly because the abdominal wall is not properly tightened during each exhalation. Women usually try to prevent potbellies; hence they tend to inhale by raising the chest cage as a whole. This manner of breathing ventilates the top of the lungs primarily and results in too much tension in the muscles of the throat, neck, and shoulder. In either case the lungs are inadequately ventilated, and the rib cage is allowed to eventually grow rigid, since its original capacity for movement is not kept up.

After having observed your habitual manner of breathing, more specifically whether you have been a "belly-breather" or a "throat-breater" up until now, re-educate yourself into being a "flank-breather" by practicing the following technique:

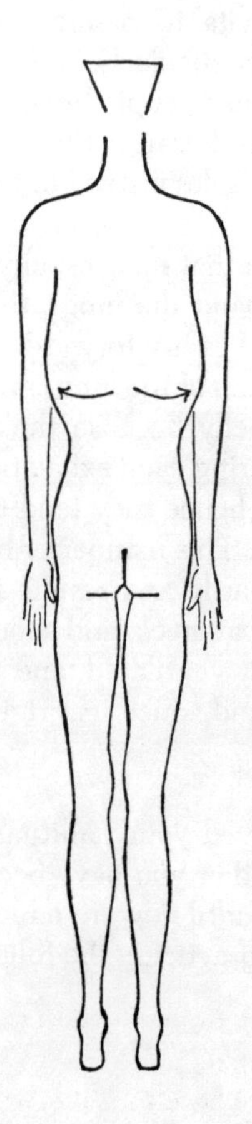

Inhale and expand your lower ribs backward and sideways, keeping the shoulder blades next to the spine and down.

Mensendieck Technique 4: EFFICIENT BREATHING

PREPARATORY POSITION: The Well-balanced Stance.
Exhale completely.

• Inhale through your nose, consciously directing the air stream toward the area immediately under your shoulder blades, and become aware of how your rib cage there starts to expand backward and sideways. Allow your lower ribs to follow this motion through until your inhalation is completed.

The air which is now filling your lungs will properly "fan out" to completely ventilate them. It is not necessary to bulge your belly, or to lift your chest toward your head. During this expansion phase of the rib cage, your abdominal muscles elongate to allow the elevation of your lower ribs.

• Exhale through your relaxed open mouth, remaining tall and upright and consciously allowing your lower ribs to sag. Exhale all the way and feel how your abdominal muscles now clearly tend to contract. Emphasize this contraction while maintaining your tall stance.

Repeat these two phases a few more times to familiarize yourself fully with the technique. When you inhale, always activate those muscles whose task it is to bring about the backward-sideways expansion of the middle part of your lungs. When you exhale, always allow your lower ribs to sag, which will bring about a firm contraction of your abdominal muscles.

If you never fully ventilated your lungs before practicing Efficient Breathing, you may initially become somewhat hyperventilated. If so, you can undo the slight dizziness by swinging your arms or taking a few steps to utilize oxygen. Then practice the Technique again, two or three consecutive times. The more your nervous system gets used to your "functional breathing," the better you will feel. *More and more you can try to breathe in this manner throughout your daily activities, until it becomes second nature. You will find yourself breathing slower than you used to, which has a definite calming effect.*

Always inhale through your nose to filter and to warm up the air before it reaches your lungs. Many people have developed the habit of narrowing the nose's air passages during inhalation. This restricts the free flow of air into their lungs. When you practice efficient breathing for the first time, watch your face close to the mirror to find out if you are also one of those who draw the sides of the nose together when breathing in. If so, close your eyes, and imagine yourself smelling the scent of your favorite perfume or flower or dish, placed directly under your nose. This will help you learn how to correctly open up your nasal passages and to inhale more freely and soundlessly.

Always exhale through your relaxed, slightly opened, mouth while practicing basic Mensendieck Routines, as if you were sighing soundlessly. The abdominal contraction from your pubic bones upward, which induces the soundless sigh, will help you gain awareness of your body's "center" to be found just above the pubic symphysis and inward. (Its exact location is stated in the first paragraph of chapter XVIII.) Exhaling through your mouth is not so strange as you may think; you always do it when you talk.

Once you have experienced how functionally correct breathing stimulates your abdominal muscles to contract effectively during exhalation, you will be able to understand why those movements which require abdominal contraction should be coordinated with exhaling.

III. RELAXATION

When a muscle contracts, it squeezes out venous blood with its metabolic waste products. In the process of this contraction, the muscle shortens, thickens, and becomes hard. When this tension is released, the muscle absorbs fresh arterial blood, with nutrients and oxygen. If the muscle is healthy, it will resume its normal shape and become softer. This explains why muscles that remain contracted for some time get in trouble. They clog up and become undernourished; they actually decrease in length and become too taut; eventually they will protest, causing you to experience pain.

Every intrusion, even upon a calm state of mind, brings about a state of "at-tention" in the whole person. When the mind is alerted, muscles tend to contract; a completely normal reaction most people are aware of. What is all too seldom realized, however, is that once the mind resolves the tension, by coping with the physical or mental or emotional impact, many muscles have to be "invited" to do the same.

The principles of how to consciously invite your muscles to release superfluous tension can be learned through the Absolute Relaxation, a perfect preparation for falling asleep. Before practicing it in your bed, familiarize yourself with this Mensendieck Technique on the floor or on your bed, where no covering will exert pressure on you. The room must be at a comfortable temperature, rather a little too warm than too cool. If you practice the Technique enough, you can eventually accomplish the total release of tension in a split second, at any time and place.

Some of our pupils, particularly pregnant women, like to have a small pillow or a rolled up towel under the knees when practicing Absolute Relaxation, because then the lower back feels more comfortable. Sleeping with such a prop under the knees is not recommended though. Rather relax and fall asleep on your side, with one thin pillow under your head and a thicker one partly tucked in between the bed and your belly, so that the latter as well as your top leg rest on it, to relieve the lower back of stress. It may feel comfortable to have the bottom arm behind your back. Avoid lying on one side more often than the other.

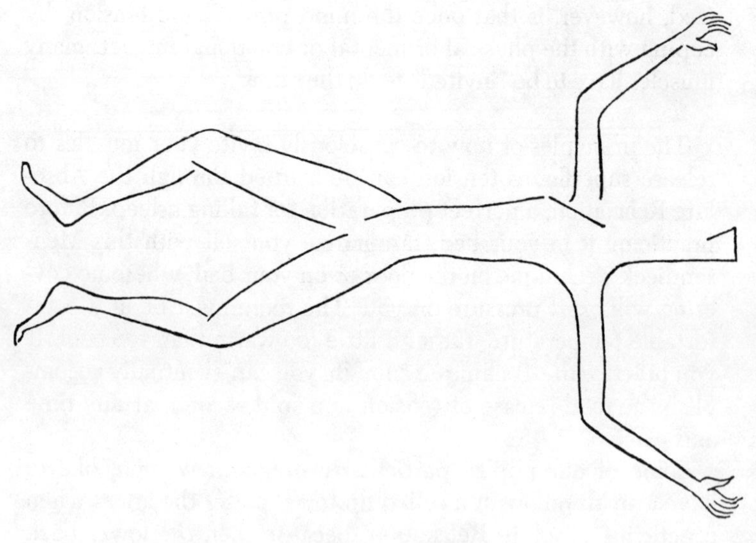

Absolute relaxation

Mensendieck Technique 5: ABSOLUTE RELAXATION

PREPARATORY POSITION: On your back; shoulders low; neck straight to crown, preferably no pillow under the head; eyes closed.

• While exhaling, tense your body by pushing your heels out, making fists of your hands, and contracting the abdomen. Become, for a moment, fully aware of the strain built up throughout the whole body.

• Slowly release the built-up strain, while inhaling.

From here on breathe normally; no specific co-ordination with the following phases is required.

• Concentrate on one of your legs; deliberately allow the tension in it to reduce even more; let it "ebb away," out of the thigh, out of the calf, out of the shin, out of the ankle, out of the toes. . . .

Your leg will now have rolled out in the hip joint, and may feel heavy, or you may not feel it at all.

• Transfer your thoughts to the other leg, and allow it to relax the thigh, the calf, the shin, the foot. . . .

This leg too has now rolled out.

• Switch your attention to one of your arms and release its superfluous tension from the shoulder toward the hand and out of the fingers. . . .

If the arm was not rolled out before, it will be so now.

• Think of your other arm, and let its tension flow out of the fingers. . . .

Allow the arm to roll out.

• Consciously relax your (tall!) neck, your jaws, your tongue, and your eyes. . . .

• Breathing has become regular, calm, and slow, relaxing your "center" inside the lower abdomen. . . .

Now you are experiencing a span of absolute rest, ideal either between activities of the day or as a preparation for a peaceful sleep to restore your vital energy for the day to come.

Rather than abruptly jerking yourself out of such a period of absolute rest, take a few moments to smoothly "stretch-tense" your muscles back to action while still lying down, and then again as soon as you get up.

At times it is good to consciously inactivate just one part of your body. For instance, after having been writing for some time, put down your pen, position your arm comfortably, then allow all tension to flow out of it. Or let your arm suspend from the shoulder, then shake it a few seconds with the hand loose, so that all the arm muscles wobble back and forth, followed by a brief period of absolute motionlessness.

Whenever your arms and/or legs have had a good workout, shake them loose in sequential order and then let them come to rest. When standing, conscious release of built-up tension in the trunk and the neck can be achieved by suspending (1) your relaxed trunk forward down in the hip joints and (2) your neck-head and arms down from the trunk. The last chapter in this book covers "Swings," the practice of which is also very relaxing to the body after it has had a workout.

In addition to absolute inactivity, two distinctly different active forms of the art of relaxation are equally vital to good health:

(1) The poise of the mind and of the body when properly centered between opposing tensions. How to achieve a poised state of mind falls beyond the scope of this book. Your body is centered when it is properly carried between opposing muscle groups of equal tone. In our introduction, we explained the importance of possessing a balanced musculature. Once your physical balance is habitually perfect you may enjoy a freedom from sensations of physical discomfort, which can help you achieve mental balance easier.

(2) The balancing of physical and mental activity. If your daily occupation requires mainly physical activity, you would be wise to compensate by pursuing something in your spare time which challenges your mind fully and enthusiastically. Your mind, like your physique, will deteriorate from non-use. If your daily occupation requires mainly mental effort, you can still use your muscles correctly through the day. In addition, if your health is good, try to engage in a sport which demands well-balanced body activity and exercises your heart and lungs, preferably in the outdoors, like hiking, mountain climbing, bicycling, sailing, skiing, or horseback riding. You may be surprised by the spontaneous solution to a problem you were struggling with at your desk, simply by letting it gestate in your relaxed mind, while your body is hard at work.

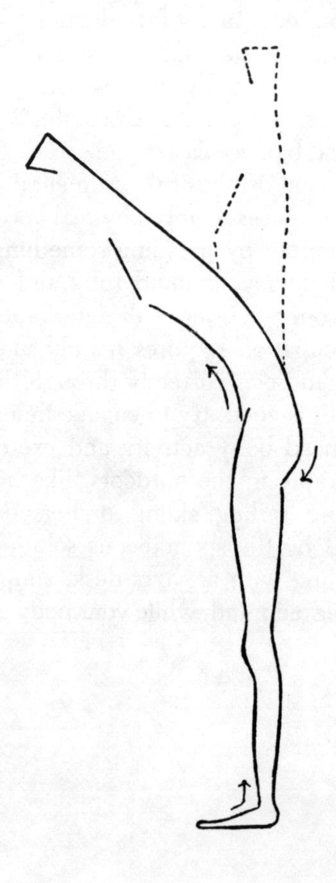

Exhale and bend your tall-to-crown spine in its five lumbar vertebrae only.

IV. BENDING YOUR BACK FORWARD

Reaching forward and down is one of man's most frequent movements. The majority of people, however, direct the body through this maneuver in a rather sloppy and potentially harmful manner. They either force their body to buckle and twist which is particularly hard on the intervertebral disks, or they incline a rigid trunk in its hip joints. The latter is quite demanding on the lower back muscles where stress already tends to accumulate during a day's work.

Normally, when we stand, the small of our back has to be held slightly concave (hollow) to countercurve the shape of our sacrum, as will be discussed in chapter XVIII. Whenever your arms have to reach forward and down, limit the bending of your back to its lower part only, curving it from concave to convex, thereby stretching your lower back muscles. *Habitually bending your back in this fashion, whenever a chance presents itself throughout your daily activities, will help prevent backaches.* The purpose of the following Routine is to acquaint you with the proper way to bend your trunk in daily life activities:

Mensendieck Technique 6: THE ROUND FORWARD TRUNK BEND

PREPARATORY POSITION: The Well-balanced Stance. Throughout the Routine, maintain your weight well forward toward the bases of your big toes, your knees slightly bent, the buttocks tight, the shoulder girdle low, your neck tall and in line with the spine.

• Inhale. (Remember to always inhale exactly as explained in Technique 4.)

• Exhale while tightening your abdominal muscles, groin to chest, and smoothly bending your tall-to-crown spine in its five lumbar vertebrae (only!), starting with the one above the sacrum, then the second, then the third, then the fourth, then the fifth, then stop.

• Inhale.

• Exhale while smoothly raising the upper part of the trunk in your lumbar vertebrae, again starting the return to their normal alignment with your lowest vertebra, then the one higher up, and so on, until your still tall-to-crown spine has resumed its ideal erect shape and place.

In the above basic Technique your feet are side by side. In your daily chores you will often walk up to whatever you intend to reach forward and down for. Then halt, after taking the last step, and maintain your weight toward the ball of your forward foot while bending and raising the upper trunk in your five lumbar vertebrae. This Step Stance enhances the beauty of your motion.

Whenever we instruct you to "keep your weight toward the balls of your feet," a firm contact of the bases of your big toes with the floor should be maintained. In the line drawings this is indicated by the little arrow in front of ankles. The base of a big toe is where the toe begins at the inner margin of the ball of the foot.

V. BENDING YOUR KNEES

The best way to experience the difference between the wrong and the right manner of bending the knees is to stand a moment and bend your knees halfway, in the way you are used to, without giving it any particular thought. You will feel that your knees can "wiggle" sideways. This causes the muscles along the back of your thighs to wobble like jelly, because of their slackness, while you sit in this uncontrolled knee bend with your derrière sticking out. Man tends to just yield to gravity whenever his body has to bend. We must learn to make our muscles bend us; the previous Technique showed you how to properly bend your back, and now you can learn how to master proper knee bending.

Mensendieck Technique 7: THE CENTRAL KNEE BEND

PREPARATORY POSITION: The Well-balanced Stance. Maintain your feet parallel throughout the Routine; only then will your legs be able to bend straight forward in ankles, knees, and hips.

• Inhale.

• Exhale while "wrapping" your corset muscles extra tight around the pelvis (abdomen up, buttocks together and down) until you become aware of increased tension in the muscles along the back of your thighs. Consciously activate these muscles to slowly bend your knees straight forward, as deep as possible, with your heels on the floor.

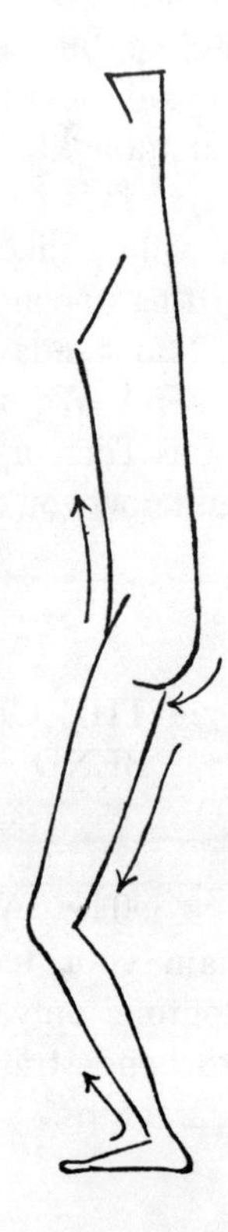

Intensify the tension in your corset muscles. Then follow through and firmly contract your hamstring muscles and those along your shins to bend in knees and ankles, while the bases of your big toes "grip" the floor.

BENDING YOUR KNEES

> Once you are sitting in the Central Knee Bend, your knees can't budge, and your lower back will be straight rather than concave. If your back is convex, you have allowed your sternum to sag; bringing it forward-up by pulling your shoulder blades to spine and down will correct this fault.

- Inhale.
- Exhale and slowly "scoop" your trunk straight upward with your buttock muscles till your body has resumed the Well-balanced Stance.

REMEMBER: A categorical imperative for advancement is to work your way through the chapters of this book in sequential order and to subsequently routinely apply **always** the basic principles of what you have learned in earlier chapters.

Many of our daily activities require a backward or a forward transfer of our body's weight, while our legs are bent. If you habitually bend your knees with your legs in Step Stance, you are prepared for those occasions. *Whenever your back has to bend together with your knees, combine Techniques 6 and 7: bend your back immediately above and your knees below an upright pelvis.*

APPLYING BASIC TECHNIQUES TO YOUR EVERYDAY ACTIVITIES

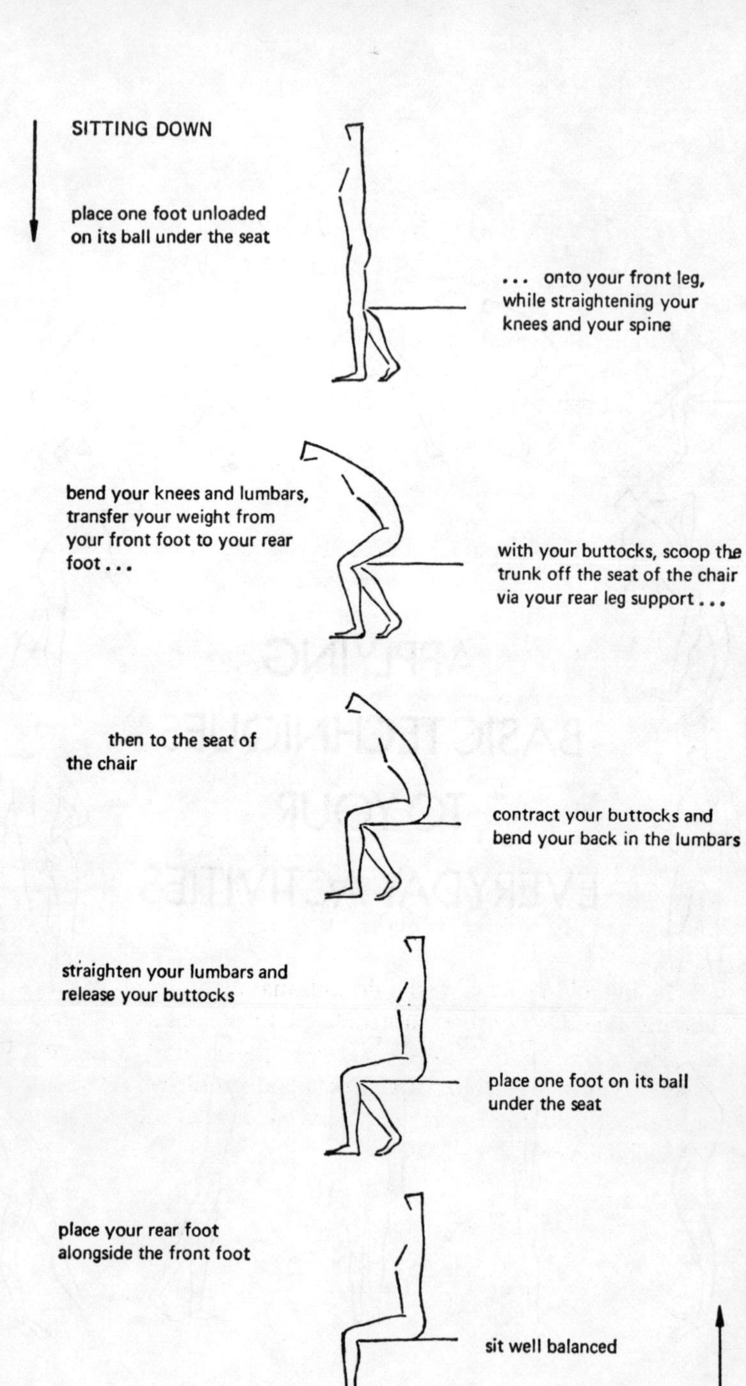
SITTING DOWN
place one foot unloaded
on its ball under the seat
. . . onto your front leg,
while straightening your
knees and your spine
bend your knees and lumbars,
transfer your weight from
your front foot to your rear
foot . . .
with your buttocks, scoop the
trunk off the seat of the chair
via your rear leg support . . .
then to the seat of
the chair
contract your buttocks and
bend your back in the lumbars
straighten your lumbars and
release your buttocks
place one foot on its ball
under the seat
place your rear foot
alongside the front foot
sit well balanced
STANDING UP

VI. SITTING DOWN—SITTING—STANDING UP

The Round Forward Trunk Bend and the Central Knee Bend should be combined in many daily routines; for instance, when you want to sit down:

1. Assume the One-legged Stance by placing one foot, unloaded, on its ball well under the seat of the chair.

2. Firmly stabilize your pelvis in its upright position with your corset muscles, and slowly bend your knees as well as your lumbar vertebrae while you transfer your weight from your front foot to your rear foot.

3. Transfer your weight from your rear foot to the seat of the chair.

4. Straighten up your back in the lumbar vertebrae and release your buttock muscles.

5. Place your backward foot alongside your forward one, parallel and at a distance of approximately two inches between balls of feet (thus three inches between heels).

The fivefold procedure should automatically result in Well-balanced Sitting, with your trunk on its sedentary (sitting) bones, located at the bottom of your pelvis, and with your spinal column aligned as it is when you stand correctly, i.e., with slightly concave lower back and neck, and a slightly convex thoracic vertebral alignment.

The Technique for standing up from a chair is exactly the reverse of sitting down:

1. Sit well balanced.

2. Place one foot on its ball well under the seat of the chair.

3. Contract your buttocks and bend your back forward in the lumbar vertebrae.

4. With your buttocks, scoop your trunk off the seat of the chair; your weight will first be carried by your backward foot before loading your forward foot.

5. Follow your weight transfer through while straightening your knees and your spine; you should now have resumed a perfect One-legged Stance.

It is essential during sitting down and standing up to maintain your neck tall to crown, exactly in line with your spine. Only then will the weight of your head adequately counterbalance and make smooth transfer of your weight feasible.

Most easy chairs preclude a foot's placement under the seat. In such a case, start out by standing with the outer side of one of your calves against the front of the chair. The foot of the leg which is farthest away from the chair is then placed backward on its ball and also in touch with the front of the chair. From here on, the technique is the same as for sitting down on a straight chair. However, you will "land" on the buttock of your front leg. After straightening your spine, gracefully pivot with the help of your hands on the seat of the chair until you are facing straight ahead; then place your feet alongside one another immediately in front of the chair. Subsequently, again with the help of your hands, slide your pelvis way back over the seat of the chair until your tailbone touches the chair's back rest, if sitting there still enables you to sit well balanced.*

To get out of such a chair: Resume the Well-balanced Sitting near the front edge of the seat of the chair. Then pivot your body back to its angular-to-chair position, while placing your feet in step position. (Forward foot must be the foot of the leg closest to the chair.) Subsequently, scoop yourself gracefully off the seat, via your backward foot onto the forward foot, as per earlier instructions for getting up out of a straight chair.

If the choice is yours: Always choose to sit on a chair with a horizontal seat level with the fold in the back of your knee. When you want to buy a new easy chair, try out several before making your selection. Sit way back, with your lumbars supported by the back rest of the chair. Relax, close your eyes, and consciously experience the sensation in your throat muscles. A skull is heavy, and when your trunk is forced to sit with its masses improperly superposed, the muscles in your throat have to "hold" your head up. Purchase the chair in which you sense the least tension in your throat. Car back rests, so that you will not be forced to slouch, ideally should support your lower back into a slightly concave shape, and allow you to sit with minimal tension in your throat muscles.

* In most easy chairs the seat is so inclined that it forces you to roll backward off your built-in "rockers" (sedentary bones). If you allow this to happen, you will slouch. Not only is sitting slouched an invitation for poor breathing and backaches, but it also impairs the nerves and the blood supply to your legs because of the disadvantageous compression of your weight. Needless to say, you would do well to avoid sitting in such a faulty posture for any extended period. If you have to sit on such an inclined seat, sit well balanced near the front edge of it. Those among you who have difficulty locating your sedentary bones, should sit on your hands, palms up, in order to clearly feel the location of these bones.

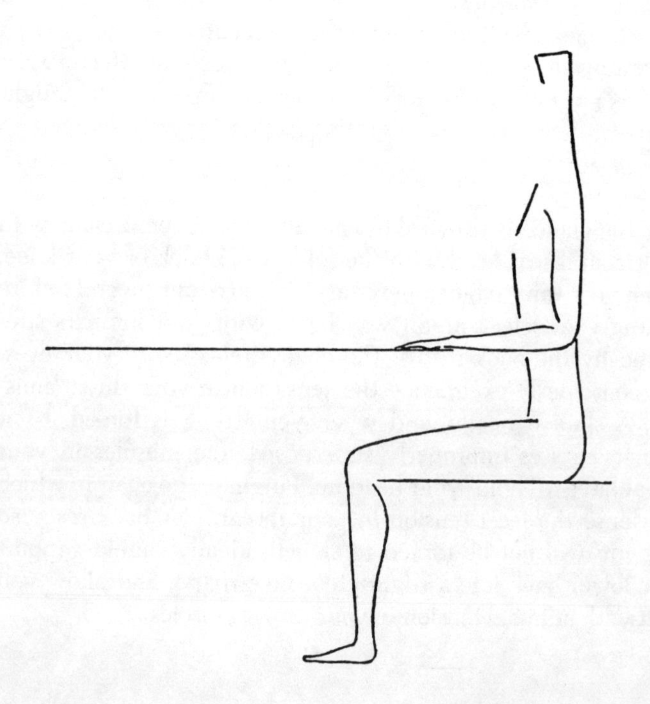

Sitting well balanced

VII. YOUR POSTURE AT THE DINING TABLE

Many Americans assume a very poor and ungraceful posture while eating. Their left forearm rests on the left thigh, except for the few instances in which the left hand has to hold the fork so that the knife can be used by the right hand. If this lopsided posture is repeated over and over again, it will ultimately result in a skeletal deformation. The spine will exhibit a twisted curve affecting the shape of the rib cage, the lower left front of the chest becomes depressed, whereas its right rear will bulge. In medical terminology this is known as a kyphoscoliosis and can readily be recognized by unequal shoulder height. Just think what this does to your muscular balance and to your lungs and breathing capacity. Visualize further how your potentially harmful dining posture compresses the stomach, which you are trying to fill up, and the esophagus through which the food has to pass before it reaches the stomach.

The European custom of *having both hands close to the edge of the table and the plate* (the fork remaining in the left hand, the knife in the right, both close above the plate until the meal has been consumed) *is conducive to sitting well balanced* as described in the previous chapter. Why not adopt this custom? All that it would really require is to learn to put the food in your mouth with the left hand. It is also much more graceful to use your knife more frequently for cutting, instead of tearing your food apart with the edge of your fork. In addition, the knife is a handy utensil to help stabilize the food which you want to put on the fork.

While discussing the effect of eating habits on posture, we may as well interject at this point that the leanness of your body is also influenced by diet. If you eat more calories than your body burns up, fat tissue will be deposited on your muscles. So here is another "balance" to be actively pursued, the one between intake and output of energy. Thus, do not expect that a reduced caloric intake alone will improve your shape; the dissolution of fat will all too often reveal the sagging and flabby condition of various muscles greatly in need of being effectively activated.

It is also worth noting that most table and desk tops are too high for the average individual. Therefore, persons who spend a lot of their time writing tend to sit with the right shoulder continually higher than the left (the left shoulder higher than the right, if you are left-handed), which will eventually also bring about kyphoscoliosis. Development of such a deformed shape can be prevented by observing two basic requirements for correct posture at a table or a desk:

The height of the seat of your chair must coincide with the fold in the back of your knees, and the height of the work surface must coincide with your elbows while the latter are straight below the shoulders and you are sitting well balanced. Many office chairs can be easily adjusted to fulfill the first requirement. The second requirement is most readily fulfilled by having the legs of the worktable or desk shortened, or by sitting with your chair on a little platform.

REMEMBER: THE BOOK IS DIVIDED IN THREE PARTS

The information presented in part one can help you acquire body-awareness and the necessary selective muscle control essential for your most basic postures and movements. In part two you are guided to immediately apply the most basic principles of correct body-management throughout all everyday activities in a manner constantly beneficial to you. Part three explains in detail how fascinatingly intricate the human body is "put together" and how perfectly it is equipped with all you need to self arrange the various body masses in their proper relationship, as well as to self release muscular discomforts, if any.

stop with inner margin of front foot alongside object,
12 to 14 inches away

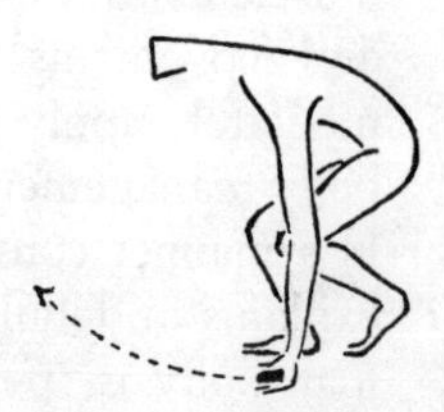

bend knees as well as lumbar vertebrae and scoop-grasp object

lift yourself forward and up with buttock muscles . . .
ready to walk on

VIII. HOW TO PICK UP A SMALL OBJECT FROM THE FLOOR

Here again, we combine the Round Forward Trunk Bend and the Central Knee Bend.

1. Walk up to the object to be picked up, and stop with the inner margin of your front foot alongside, but approximately 12 to 14 inches away from, the object.

If you have walked correctly, you will halt with your body's weight carried toward the ball of your front foot and with the ball of your rear foot still in touch with the floor.

2. Stabilize your pelvis firmly with your corset muscles, and slowly bend your knees as well as your lumbar vertebrae till you can "scoop-grasp" the object with the hand on the side of your rear leg.

Keeping your neck tall, in line with your spine, is conducive to moving your arm in its shoulder joint, while your shoulder blades are still in correct position, i.e., low and next to your spinal column. All this will help prevent your bending any of the vertebrae above your lumbar ones.

3. Intensify the tension in your buttock muscles and allow them to lift you forward and up, out of your knee bend, while you straighten your upper trunk in the lumbar vertebrae.

If you follow this lifting of yourself through, all the way, finishing with your body diagonally tall to sternum, as well as your neck tall to the crown of your head, you will automatically find yourself walking again. Tight buttocks provide forward propelling power to the body when it is properly aligned.

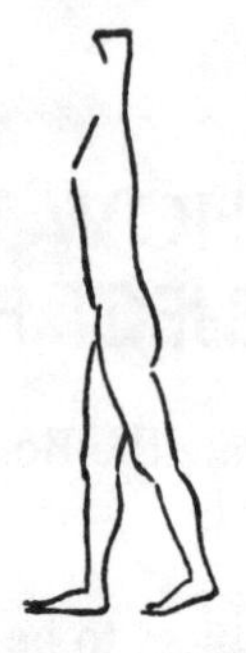

in step position, body weight on both feet, tighten your corset muscles

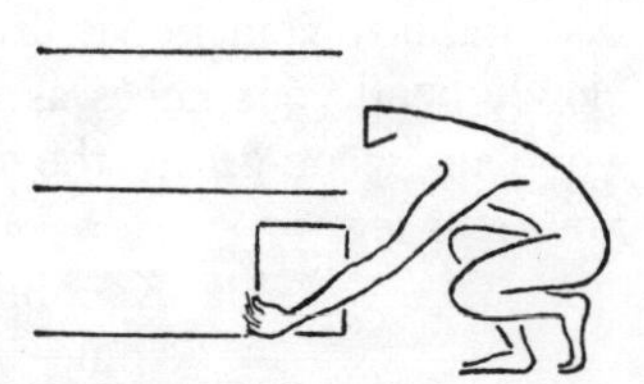

to protect your back, exhale while securing a firm abdominal contraction which allows your body's power source (center) to cope with the lifting of the heavy object

maintain correct postural alignment and carry load with your corset muscles close to your center

IX. THE LIFTING AND CARRYING OF A HEAVY LOAD

In principle the technique for lifting a heavy load is the same as for the picking up of a small object. However, because you will need both hands and the muscles in both legs, place yourself in step position directly in front of the object, with your body weight equally distributed over both feet.

In order to protect your back during the effort, exhale through your relaxed open mouth to assure a firm abdominal contraction while lifting the load. *It should feel as if your corset muscles, rather than your arm muscles, are doing the lifting.* Maintain your neck tall and your shoulder girdle low.

If you have to get an object from way down, for instance out of a low drawer or the bottom shelf of a low cabinet, bend your trunk and your knees in the proper way until you are sitting on the heel of your rear foot. Subsequently, lift yourself and the object with your corset muscles, while exhaling.

Avoid carrying heavy objects, particularly over a considerable distance. But if you have to, carry the load close to the center of your body, while maintaining correct alignment. Again, firm activation of your abdominal muscles will help to protect your back from harm; their tension keeps the upper back from inclining backward in order to counterbalance the weight you are carrying in front. Mechanical dolls would have to counterweigh with the upper back, but you have muscles, hence, you don't have to. Pregnant women should heed these words in the later months of their pregnancy. Welcome the additional challenge to your abdominal muscles; use them; it will keep them strong, which will prevent much discomfort in your back and greatly assist natural childbirth.

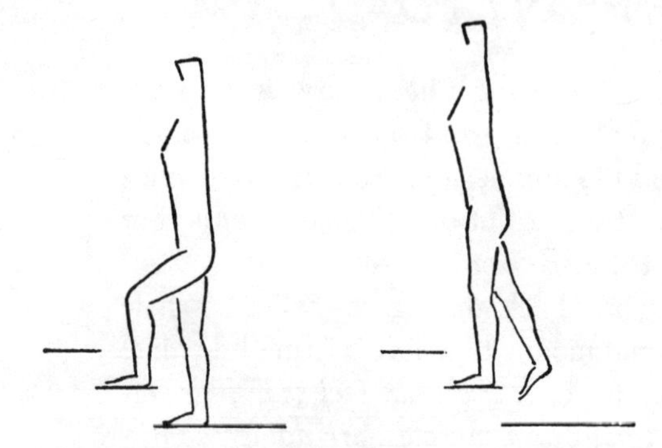

climb on your front lower leg . . .
. . . follow your sternum forward and up

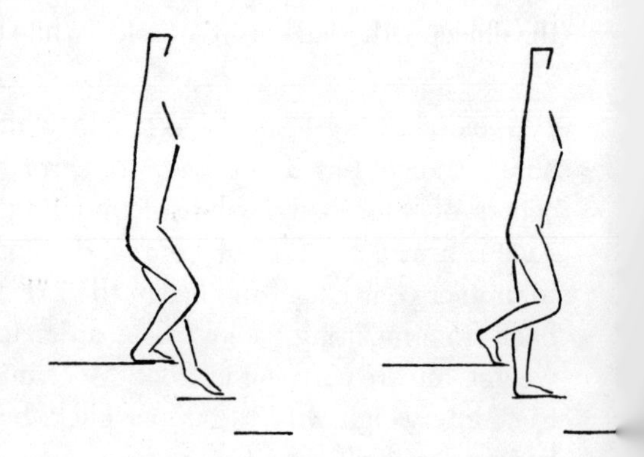

"central knee bend" your rear leg . . .
. . . until front foot lands loaded

X. GOING UP AND DOWN STAIRS

This activity provides you with ample opportunity to practice the basic principles of the Central Knee Bend. *Going upstairs, properly, is most demanding for your forward leg, and going downstairs requires the main action in your backward leg.* Concentrate, during the former, on following your sternum forward-up while climbing on your front lower leg; during the latter, concentrate on placing your straight trunk on an imaginary seat.

The easiest way to start practicing the correct manner of climbing up a flight of stairs is by first assuming the Well-balanced Stance immediately in front of the staircase. Make particularly sure that your weight is equally distributed over your heels and the balls of your feet so that you are thus standing in a slightly diagonal plane.

1. Place one foot on the bottom step, parallel to the foot of your rear leg. Become conscious of your front lower leg and think of it as being a pole on which to climb.

2. Lift yourself onto that pole by increasing the tension in your buttocks as well as in front of your forward ankle, knee, and thigh. Follow your sternum diagonally upward until your forward leg is straight in knee and hip joint.

   Your straight rear leg has been carried along rather passively during your climbing. The backward foot should be hardly used for pushing you on your imaginary pole.

After having fully completed your first step-climb, bring your rear leg forward while bending its knee, and place its foot on the next step of the staircase. Carefully climb on its lower leg. And so on. Once you have reached the stairhead, make an about-face and resume the Well-balanced Stance close to the edge, as a preparation for learning how to descend stairs correctly. Then transfer your weight to one leg by moving your upright pelvis sideways, so that the upper part of your trunk will not lean that way.

1. Increase the tension in your corset muscles and make them bend your loaded leg while bringing your unloaded leg forward in its hip joint till its foot is above the top of the stairs.
2. Increase the knee bend in your standing leg to land your forward foot on the step with part of your body's weight in it. Follow through and transfer your total weight.

Subsequently bring your backward leg forward while straightening its knee, to bring its foot above the next step down, and again execute the Central Knee Bend in your standing leg until the foot of your extended leg has landed and carries you. And so on.

Once you have mastered the proper Technique, applying its basic principles in a more fluent sequence will automatically result in natural momentum. The well-controlled muscular coordination will have made handrails superfluous for you, because of your superb balance. It is, however, always better to let one hand slide along the rail so that it can sustain you in an emergency. As applies to all arm movements under shoulder level at all times, maintain your shoulder girdle low (your neck tall to crown), with the shoulder blades alongside your spine, while keeping your hand in light touch with the handrail.

To be applauded for somehow accomplishing some kind of balanced stance is timely for toddlers in the process of exploring the two-legged upright position of man. However, the initially haphazard superpositioning of body masses should never be allowed to become a set pattern for the remainder of an individual's life.

KNEELING DOWN

place one foot, unloaded, on its ball behind you

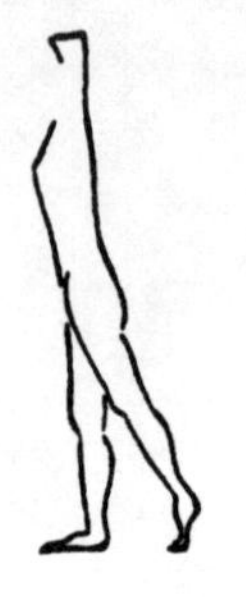

lift trunk forward and up with your buttocks while exhaling

"central knee bend" till rear knee lands on floor

place rear foot on its toes

extend rear foot

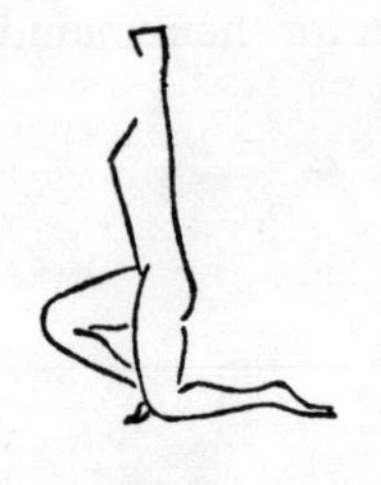

place foot of unloaded leg next to loaded knee

place front lower leg alongside other lower leg

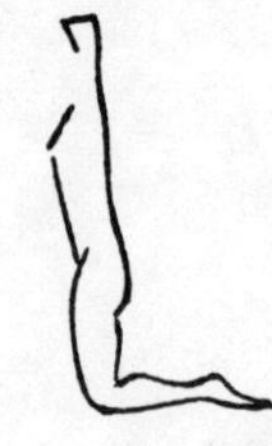

transfer your weight to one leg

distribute your weight over both legs

GETTING UP

XI. KNEELING DOWN AND GETTING UP

Start out from the Well-balanced Stance, then transfer your weight to one foot by moving your pelvis sideways. Be sure to keep the pelvis frontal and upright.

1. Place the unloaded foot backward, on its ball, and at a distance of at least ten inches. Your feet should remain parallel throughout the Routine.
2. Increase the tension in your corset muscles and make them bend your knees until your rear knee lands gently on the floor next to your front foot.
3. Extend your rear foot so that its instep and extended toes are on the floor.
4. Bring your front lower leg straight backward and place it alongside your other lower leg. Distribute your weight over both legs.

Getting up from the floor is the reverse procedure:

1. Transfer your weight to one knee by moving your upright pelvis sideways.
2. Raise the unloaded bent leg straight forward until you can place its foot next to your rear knee.
3. Bend your rear foot in ankle and toes until that foot is supported by its toes.
4. Firmly tighten the muscles in your legs, while lifting your trunk diagonally forward-up with your buttocks, until your knees are straight. Considering your front lower leg a pole on

which to climb will help you scoop your trunk in line with it, preferably while exhaling because of the considerable effort required.

Correctly, you should now stand with your weight toward your front ball of foot, your body tall to sternum, your neck tall to crown. Bringing your rear leg alongside your standing leg by increasing the tension in your corset muscles, and then distributing your weight over both feet, should result in the Well-balanced Stance.

NOTE: To acquire a balanced musculature, you must always demand equal work from both sides of your body. It can be very revealing to become aware which leg you tend to favor. If you have noticeably more difficulty with one particular leg in the forward position, place that one in the front most of the time until its strength equals that in your other leg. Regularly alternate between the two legs from then on. People with ailing knees should always have the better knee in front and may need the help of their hands on the forward thigh.

The reasons why correct body-maneuvering does not come naturally in man, and therefore has to be learned like riding a bicycle or driving a car, are extensively explained in the Introduction to this book dedicated to all people interested in achieving skilled manipulation of their body-mobile. This skill, well applied, will bring about perfection in the individual's carriage and shape, as well as enhance his well-being.

LOWER BODY TO FLOOR

kneel and raise arms

pull your arms down

lift trunk with buttocks
and thigh muscles till
you stand on knees

"central knee bend" to
seat trunk on heels

scoop up trunk and
seat it on your heels

sit on one side of
both heels

bring both feet alongside
one of your hips

straighten lower legs in
line with upper legs and
stabilize legs by
raising your feet in ankles

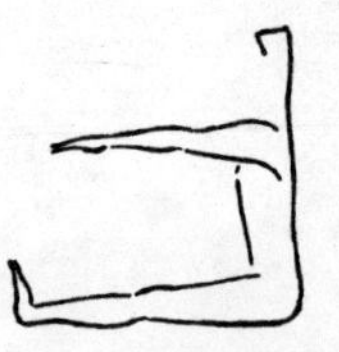

sit up while exhaling and
straighten your back fully

lower your back from tail
to crown while exhaling

secure your feet in
vertical, then inhale

GETTING UP

XII. HOW TO LOWER YOUR BODY ALL THE WAY TO THE FLOOR

Occasionally people have to sit or want to lie on the floor. Hence, you may as well learn how to correctly *"fold" yourself in and out of these positions in a graceful and easy manner while exercising your muscles.* Each phase of the following two Mensendieck Routines is an exercise in its own right and can be repeatedly practiced separately, to recondition muscles where you feel this is needed. First, we will discuss how to lower your body until you lie supine (on your back) on the floor:

1. Kneel down, as explained in chapter XI. Subsequently, while keeping your shoulder girdle low and its blades alongside your spine, slowly elevate your arms forward in the shoulder joints to shoulder level.

2. Tuck your buttocks firmly under and slowly lower your erect trunk till you sit on your heels.

3. Seat yourself on the floor on one side of both heels, while you bend and rotate your upper trunk toward the opposite direction, in order to adequately counterweigh for balance.

4. Extend your legs, by swinging your lower legs around, above the floor (which will make you roll from sitting on one buttock till you sit on both of them), and straighten up your spine immediately after your extended legs land gently on the floor. Subsequently, raise both feet in the ankle joints, keeping your toes in line with the insteps of your feet. Your straight legs are now stable.

5. With your arms still horizontal, curve your spine backward and gently lower your back to the floor, first your sacrum, then the lumbar vertebrae, then the thoracic vertebrae, then

your neck, and finally your head at the same time that your arms land on the floor. While so uncoiling your spine, make it a point to always exhale because of the special demand on your abdominal muscles.

Getting up from this supine position is the reversal of the lying-down procedure:

1. Stabilize your legs by securing your feet in the vertical. Thinking of pushing your heels away, rather than pulling your toes up, helps to maintain the latter in line with the insteps of your feet.

2. While exhaling, firmly contract your abdominal muscles and lift your back off the floor, starting with your head and horizontal arms. Once you are in sitting position, be sure to subsequently straighten up your spine from tail to crown. You have now accomplished the one and only beneficial "sit up."

3. Relax your feet in the ankles and swing your lower legs around till both feet are on one side of your now sideward-tilted pelvis, and you are resting on one buttock.

4. Scoop up your trunk (with the buttock muscle on which you were sitting) and seat it, upright, on your heels.

5. Lift your erect trunk by contracting the buttocks and thigh muscles, until you stand on your knees. Slowly lower your arms in their shoulder joints (girdle low, blades snug to spine). Stand up from your knees, as instructed in chapter XI.

NOTE: When you practice one phase and its reversal a couple of times consecutively, to recondition specific muscle groups, co-ordinate this with the efficient breathing Technique as follows: With the exception of the lifting of the arms, during which you should inhale, exhale during each on-your-way-down phase, as well as during each on-your-way-up phase. Consequently, inhale in between each of these movements which require the co-operation of your abdominal muscles, both ways.

The third part of this book is devoted to some more advanced Mensendieck. At one point, the prone (on your belly) preparatory position will be required. The easiest way to get into prone position on the floor is by rolling over, once you have lowered your body into the supine position, as instructed above. There is however a more challenging, direct way to get down on your belly, which some of the fittest ones among you may like to try. One word of caution though: The following Routine and its reversal are definitely contraindicated for patients with deteriorated lumbar disks.

1. Stand well balanced.

2. Bend your head forward, then your neck, then the thoracic vertebrae and lumbars, eventually your pelvis in its hip joints, until the hands are on the floor. Remain standing toward the balls of your feet, knees straight, during this phase. (It is always preferable to have your knees slightly bent, rather than locked in hyperextension). From here on, maintain neck tall to crown throughout the Routine and its reversal.

3. Straighten out your body by walking straight forward over the floor with your hands (elbows extended) until your feet stand on their toes, i.e., until the balls of feet have all but cleared the floor.

4. Slowly lower the knees onto the floor and extend your feet (insteps on floor).

5. With your elbows still straight (if, or as long as, your spine allows that), slowly lower your pelvis onto the floor.

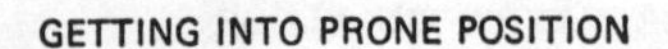

GETTING INTO PRONE POSITION

stand well balanced

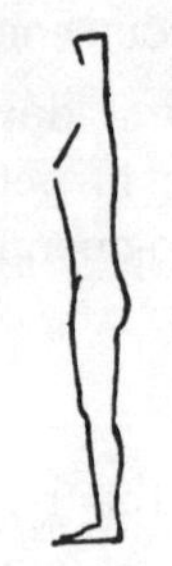

raise trunk, pelvis first, crown last

bend, head first, and place hands on floor

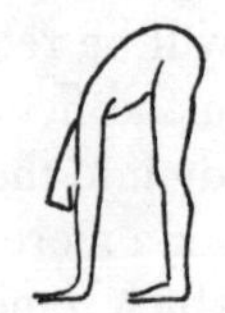

walk with your hands to your feet

walk forward on hands until your feet stand on toes

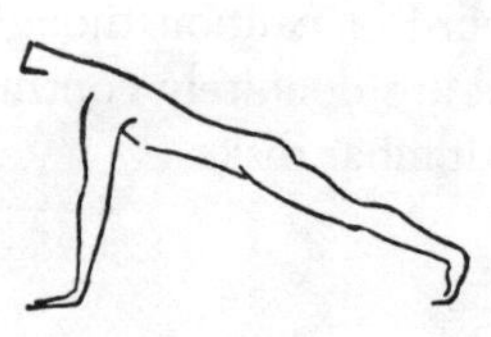

place feet on toes, then lift your knees

lower knees to floor and extend your feet

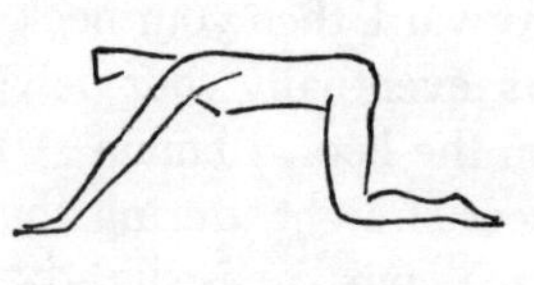

lift pelvis till it is above your knees

lower your pelvis to floor

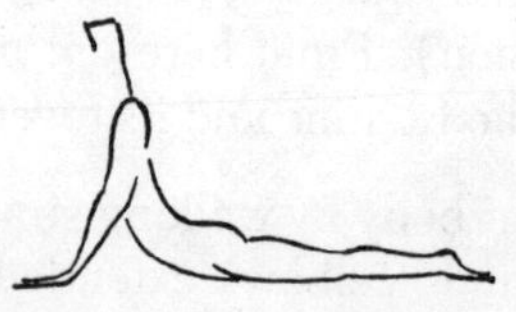

straighten your elbows

lower your upper trunk

raise neck to lift forehead off floor

lower neck till your head rests on forehead

GETTING UP

6. Slowly bend your elbows to bring your shoulders above your hands with your tummy on the floor.

7. Slowly bend your tall neck until your forehead lands on the floor.

Getting up from this prone position is the reversal of the lying-down procedure:

1. Secure your hands below your shoulders on the floor, hands parallel to your nose.

2. Raise your neck to lift your forehead off the floor (and to get your neck tall to crown again!).

3. Slowly straighten your elbows all the way (if your spine allows that). You have now accomplished the best "push-up" yet. (Frequently practicing this phase and its reversal will help clear up accumulated strain in lower back muscles and will correct flabby upper arms.)

4. Lift your pelvis till it is above your knees.

5. Move in ankles and toes until your feet are supported by their toes, and lift your knees off the floor until your body is straight.

6. Walk with your hands toward your feet, until your body is in jackknife position (head down!) and the soles of your feet are on the floor.

7. Straighten up your body by first raising the pelvis in its hip joints with your buttock muscles, and then stack your vertebrae from sacrum to the crown of your head until your spinal column is again in perfect upright alignment with your legs.

SIT-UP PER LEG PUMP

push heels away to place
feet in vertical

straighten your leg in
hip and knee

lift one leg while keeping
its lower leg horizontal

unlock your hands

firmly interlock hands
below knee fold

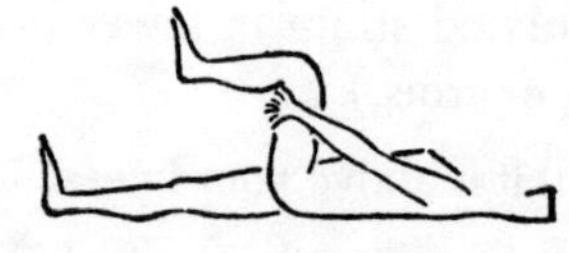

lower your neck
and head

start exhaling and raise
your head . . . decide on
pushing your heel away next
while you suspend from your arms,
trunk relaxed

keep trunk relaxed and
slowly lift leg with its
lower leg horizontal

follow through exhalation and
firmly "pump" your heel away
horizontally, which will make
you sit up with a relaxed tummy

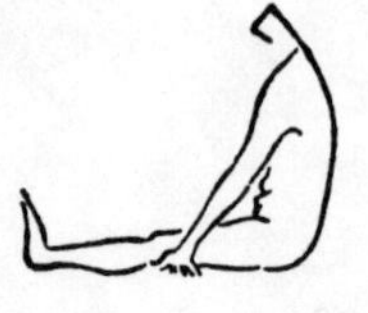

with your feet in vertical,
suspend from your arms and knee

LYING DOWN PER LEG PUMP

XIII. THE LEG PUMP SIT-UP AFTER SURGERY, CHILDBIRTH, ETC.

Whenever the sit-up Routine as described in chapter XII is contraindicated, as for instance after abdominal surgery, the Leg Pump will allow you to assume a relaxed sitting position with the least discomfort. In most cases, this Routine can be practiced before you go into the hospital, so that you can execute it with confidence when needed. *All you have to do is to learn to utilize the muscles in your legs instead of those in your tummy:*

1. Place your feet perpendicular to the legs by firmly pushing your heels away from you (rather than pulling up your toes!). Maintain your feet vertical throughout the Routine.

2. Slowly lift one leg in the hip joint. Simultaneously bend the knee and maintain the lower leg parallel to the bed. Maintain the lower leg horizontal throughout the Routine.

3. Interlock your ten fingers immediately below the fold in the back of the raised knee.

Very shortly, your completely relaxed trunk (chin near chest) will be suspended from the straight arms, so be sure that your fingers are tightly interlocked as close under the knee as possible.

4. Start exhaling and raise your head chin-to-chest. Decide on pushing your top heel away next, while suspending with your trunk from your straight arms.

5. Continue to exhale while slowly but firmly straightening the elevated leg (foot vertical!) forward and down . . . without bending your arms at all. By the time your moving leg has straightened, it lands on the bed and you will have assumed a relaxed sitting position.

To lie down, follow the same procedure in reverse:

1. With your feet in the vertical, suspend from your arms and one knee, trunk relaxed and your chin near chest.

2. Slowly lift one leg, maintaining its lower leg horizontal, to lay your trunk down.

3. Lower your neck and head.

4. Unlock your hands.

5. Straighten your elevated leg in hip and knee joints. Then relax your feet.

One of the greatest endowments of the human body is its power to perfectly perform millions of intricate movements.

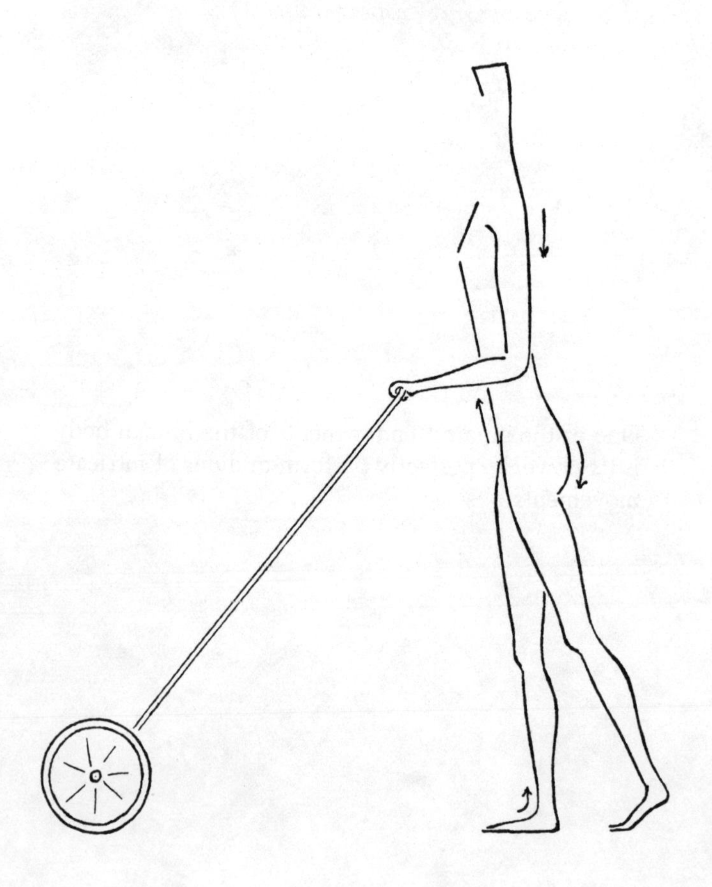

Your corset muscles should push, whatever the object.

XIV. PUSHING

You can achieve efficient pushing if you *walk the Mensendieck way*, keeping your body's weight well ahead of your parallel feet, *while emphasizing the tension in your corset muscles*. The shoulder girdle must be carried low, shoulder blades alongside spinal column, elbows alongside rib cage.

Pushing correctly feels as if your abdomen and buttocks are doing it, rather than your legs and arms. If you are aware of extra tension in your back, you are really inviting trouble.

If the object to be pushed is heavy and not mounted on wheels, move it only about a foot at a time.

Inhale while pausing and securing your hold on the object; exhale while securing the muscles' hold on your pelvis, and your subsequent pushing effort. Discontinue pushing once you start to inhale again.

XV. HOW TO AVOID FLABBY UPPER ARMS

While the bending of elbows and fingers is required time and again throughout our waking hours, we seldom wholly straighten them. This results in an unbalanced muscle tone between the flexors and the extensors of our arms.* The fact that we all have far less tone in the latter than in the former can clearly be observed when the arms hang relaxed alongside the body.

We particularly neglect using the extensors of our elbows, which run along the back of our upper arms. That's why they tend to become flabby. To reverse this tendency the following principles have to be applied routinely:

1. Execute all arm movements which take place below shoulder level with your shoulder girdle stationary in correct position, i.e., with the shoulder blades low alongside the spinal column and snug against the rib cage.

2. Always initiate your arm motions at the shoulder in order to correctly engage the muscles which run from the shoulder girdle and the trunk to your arms.

If the upper arm also takes part in the action, start your arm's movement in the shoulder joint; if the action is with the forearm and hand, stabilize your upper arm with the muscles around the shoulder joint, particularly those along the back of your armpit.

3. Fully straighten your arms in elbows and fingers frequently, because none of our daily tasks ever requires this.

* Flexors are muscles which serve to bend adjoining parts of the skeleton relative to one another; extensors are muscles which serve to straighten such parts.

4. Rotate straight arms in the shoulder joints, as often and as firmly as you can, to make up for the fact that tasks which require a twisting with your hand take place in the forearm only. (Remember: Your shoulder girdle may not move.)

5. Your shoulder blade has to leave its position alongside the spinal column as soon as an arm is raised above shoulder level; make a habit of allowing this "reluctantly," that is, against conscious resistance. After having reached high, always pull your arm down by drawing the shoulder blade low toward your spinal column first.

XVI. A WORD ABOUT OCCUPATIONAL POSTURES

Many occupations require a faulty postural alignment to a greater or lesser extent. Real harm to the body results if it is also carried faultily during the remaining hours of the day, which unfortunately is often the case. Surgeons, for instance, are notoriously hunchedbacked; dentists high-shouldered on one side; ballet dancers flat-footed; hairdressers bulge-bellied; fashion models rigid-backed; competitive swimmers stooped-shouldered . . . to mention just a few typical "occupational" postures.

Although the real need for faulty carriage during the actual performance of certain duties is debatable, people will no doubt continue to indulge in it while concentrating fully on a difficult task at hand. However, once such a task is accomplished, these men and women should be overly conscientious in normally carrying their body masses perfectly balanced.

Correct posture does not necessarily have to be mysteriously elusive.

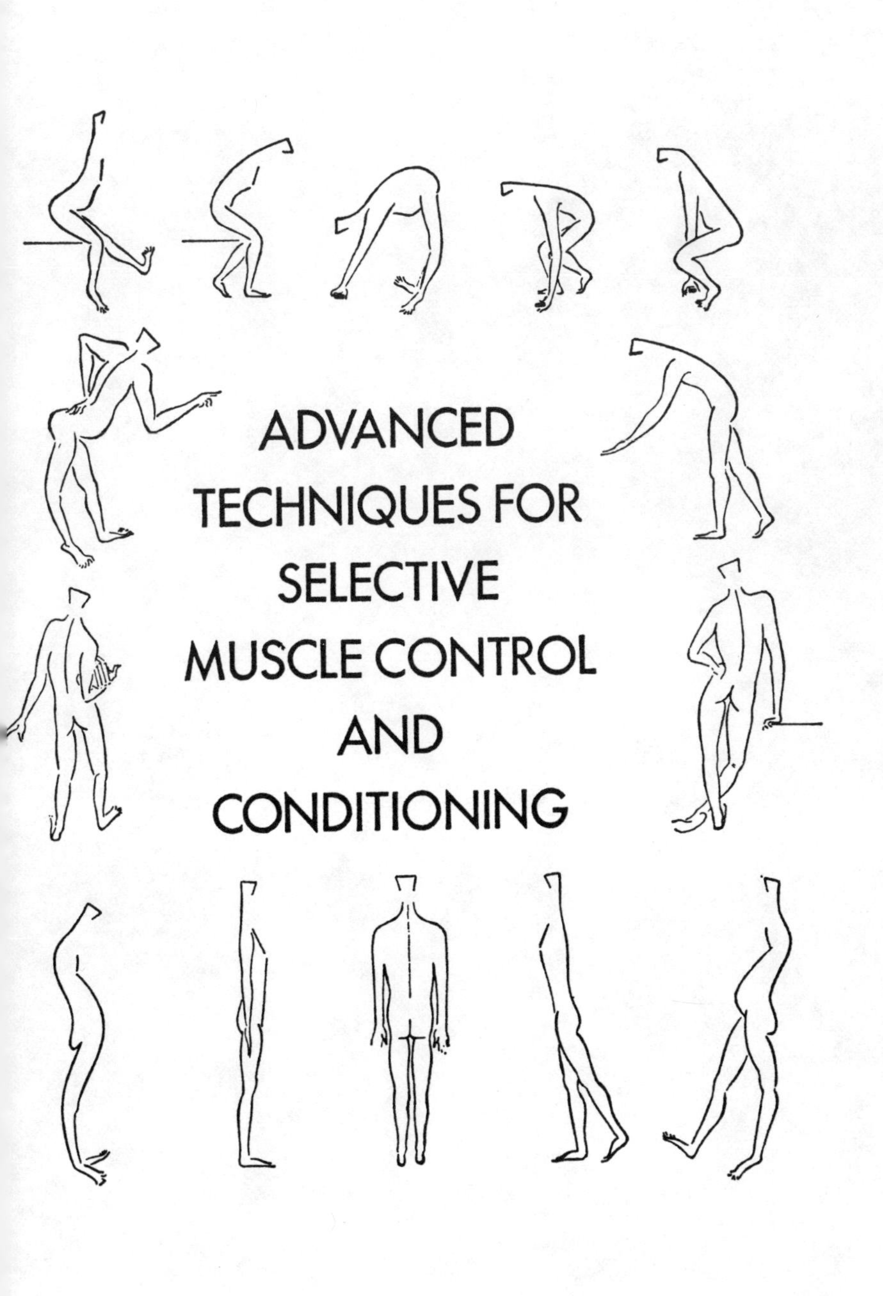

ADVANCED TECHNIQUES FOR SELECTIVE MUSCLE CONTROL AND CONDITIONING

XVII. GET ACQUAINTED WITH YOUR FEET

The foot is a complex unit, composed of twenty-six bones and numerous muscles. Its main function is weight-bearing. This function is called to task whenever an individual is standing stationary or moving about. Miraculously, the construction of the foot enables it to meet the demands of constantly changing conditions. It has sufficient solidity to carry the weight of your body, yet it possesses adequate capacity for pliancy to adapt itself to walking over all kinds of terrain. The foot muscles must be strong enough to absorb the impact of the body's weight with each step as well as resilient enough to time and again bring about the foot's resumption of its ideal shape when unloaded.

Because of its complexity, the foot is subject to many disorders which can make standing and walking miserable. If you are already experiencing foot discomfort, you will benefit greatly from the Routines discussed in this chapter. If your feet are in excellent condition, these Routines will help you to keep them in tiptop shape. Practice them whenever you have a chance, for instance while taking your bath! Many of our pupils find that very relaxing.

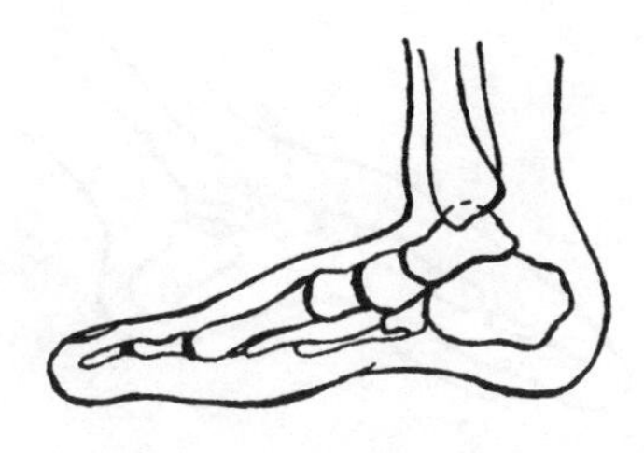

Medial view of the bones of the right foot

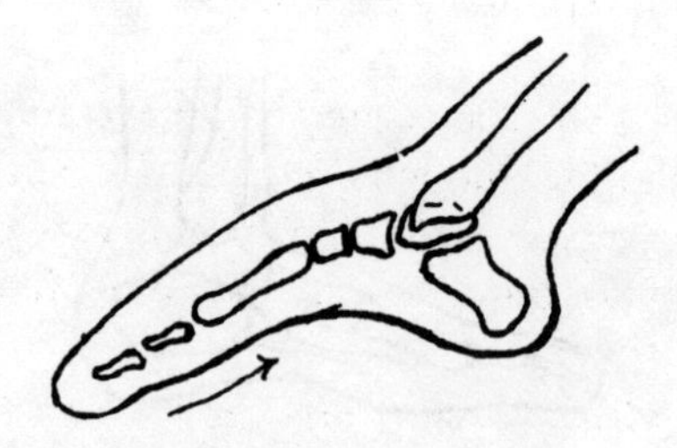

Cupping the ball of your foot

Mensendieck Technique 8: YOUR TOES AND YOUR STIRRUP MUSCLES

A. Cupping the Foot
B. Outer Margin Grip

A.

PREPARATORY POSITION: Sit well balanced on a stool. Cross one thigh over the other. The foot of the supporting leg should be on the floor, with its heel straight below the knee. Observe the relaxed foot of the leg which crosses the supporting leg and note how its toes tend to be in a slightly raised position relative to the instep. This corresponds with having the ball of that foot somewhat protruded. (This happens because people use the muscles which pull the toes up much more than those which pull them down.)

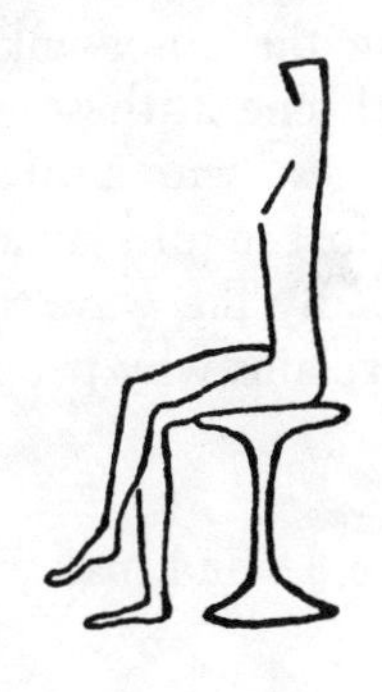

• Straighten your toes in line with the instep and notice the accompanying tension of the muscles in the ball of your foot as well as the resultant hollow shaping of this ball.

• Gently release the muscles in the ball of your foot, which will result in your toes springing back to their previous position.

Repeat these two phases until you have fully familiarized yourself with the Technique of "cupping" the ball of your foot slowly and consciously. Then cross your legs the other way around and practice the same with your other foot.

This Mensendieck Routine aims at the re-establishment of the proper balance between the flexor and extensor muscles of your toes by strengthening the usually neglected flexors.

B.

PREPARATORY POSITION: The same as for Technique 8A. Observe the relaxed foot of the leg which crosses the supporting leg and note that the base of the big toe is closer to your inner ankle than the base of the little toe is to your outer ankle. (This happens because people generally use the muscles along the front of the inner ankle much more than those along the outer ankle. This results in unequal tone in these all-important "stirrup muscles," with a subsequent tendency of the foot to turn inward when it hangs relaxed. By the way, this muscular imbalance explains why people "twist an ankle" outward.)

- Cup the ball of your foot as in Technique 8A and maintain your toes in this position.
- Contract the muscles which run along the outside of the outer ankle and raise this margin of the foot closer to the outer ankle, while maintaining the toes stationary. Observe how now the distance between your inner ankle and the base of your big toe has increased.
- Release the grip on the outer margin of your foot, while maintaining your toes stationary.
- Release "the cup" in your foot.

Repeat these four phases, slowly and consciously before doing the same with your other foot. If phases two and three are correctly executed, muscle tension can be felt under the ball of the foot and along the outside of your lower leg where the muscles to the outer foot margin originate.

This Mensendieck Routine aims at balancing your stirrup muscles, as well as at strengthening the muscles in the balls of your feet.

The responsibility for stabilizing you at the ankles is among others shared by the tibialis muscle and the peroneus longus muscle; referred to in Mensendieck as your stirrup muscles. The tibialis originates at the front of the lower leg and runs along the front of the inner ankle toward the midpoint of the inner foot margin. The peroneus longus originates at the outside of the lower leg and runs along the outer ankle, then wraps around the outer foot margin to continue its course under the foot sole toward the midpoint of the inner foot margin. Activating the tibialis only, when the foot can freely move, will pull the inner margin of the foot closer to the lower leg. Activating the peroneus longus only, when the foot can freely move, will pull the outer margin of the foot closer to the lower leg. These movements are known as supination and pronation respectively. Simultaneous activation of tibialis and peroneus, when the foot can freely move, will elevate the foot in the ankle (dorsiflexion).

While standing, the activation of your stirrup muscles must secure correct carriage of the longitudinal arches of your feet relative to your lower legs, thus preventing flat feet, as well as your lower legs relative to the feet, thus preventing hyperextension of the knees. Hence, whenever a Routine requires that you keep your knees slightly bent while standing, from now on you should do so by activating your stirrup muscles in addition to your corset muscles to maintain your lower legs correctly related to your feet and vice versa. Simultaneous activation of "stirrup and corset" muscles is the most basic requirement for the construction of perfect posture.

When you succeed in activating your stirrup muscles, the lower (tendon) part of the tibialis muscle will become clearly

right lower leg

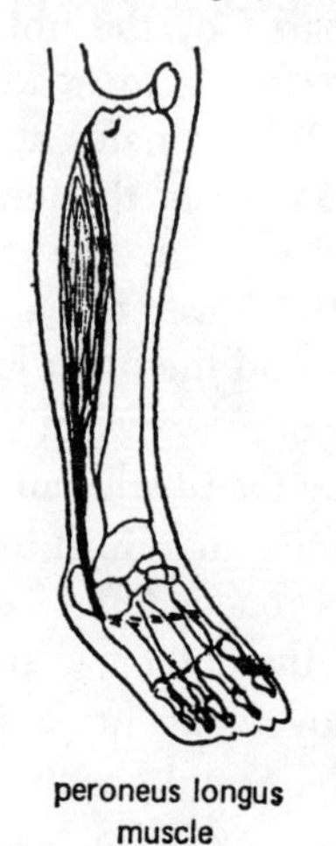

The principal stirrup muscles

visible in front of your inner ankle; it pops off while taking the short cut. Few people experience difficulty in activating the stirrup muscles when they are sitting and the feet can freely move relative to the legs, as will be instructed in Techniques 9 through 12. All people, at first, experience difficulty in activating their all-too-lazy stirrup muscles while standing. What will help you to get a hold of them is to carefully execute the Central Knee Bend Technique (7), not only by wrapping your corset muscles tighter (abdomen up, buttocks together and down), thereby inducing tension in the muscles along the back of your thighs, but also by bringing about a tightening of the muscles along your shins and outer ankles until both your tibiales "pop off." Then try to, slowly and firmly, straighten the knees till the Well-balanced Stance is resumed . . . without releasing your stirrup muscles at any time. Once you are again standing correctly erect: You will now feel that the bases of your little toes hardly contact the floor, and that you in fact now cup (narrow) the balls of your feet by drawing the bases of your little toes closer to the bases of your big toes. If you do not experience the non-loading of the bases of your little toes as yet, there is no reason to despair. At the conclusion of your practice of Technique 14, you too will succeed.

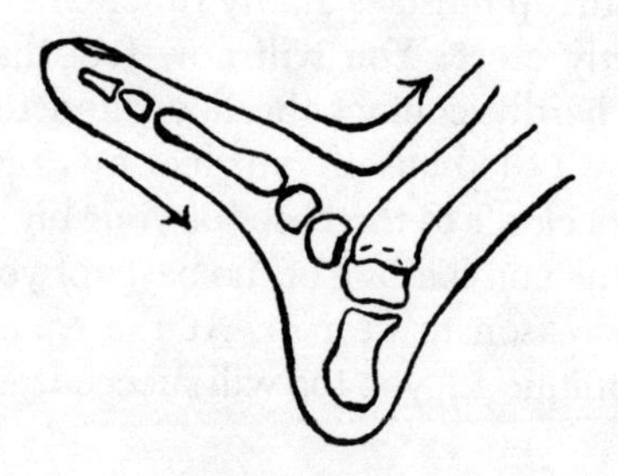

The well-balanced foot position.

Mensendieck Technique 9: THE WELL-BALANCED FOOT POSITION

PREPARATORY POSITION: The same as for Technique 8A.

• Cup the ball of your unloaded foot, to extend your toes in line with the instep.

• Get some grip on the outer margin of your foot while maintaining your toes stationary. You have now activated your peroneus muscle.

• Slowly raise your foot all the way in the ankle joint, while maintaining the cup in the ball of the foot and some grip on its outer margin. You have now also activated your tibialis muscle.

The resultant Well-balanced Foot Position is the preparatory position for basic Mensendieck foot exercises as well as the position in which the foot is carried when a leg is being moved during Mensendieck training.

• Slowly release the muscles along the front of your ankle, while maintaining your toes in line with the instep.

• Release the muscles toward the outer margin of your foot.

• Release the cup in your foot.

Repeat these six phases a few more times, slowly and consciously, before doing the same with your other foot.

This Mensendieck Routine aims at strengthening your feet and ankles.

Mensendieck Technique 10: THE TOE SPREAD

PREPARATORY POSITION: Sit on the floor with straight legs and back, neck tall, shoulders low. If you need the support of your arms, place your fists (palms backward) at a little distance behind you on the floor, to keep your elbows from hyperextending. (If the above preparatory position is not possible for you as yet, assume the preparatory position for Technique 8 and work with one foot at a time.)

• Assume the Well-balanced Foot Position simultaneously with both feet, as explained in Technique 9. (Briefly: Toes extended in line with instep, cupped ball of foot, some grip on outer foot margin, foot raised in ankle joint.)

• Spread your toes, while keeping them in line with the instep.

• Close your toes tightly together, while keeping them in line with the instep.

• Release all leg and foot muscles and allow the feet to resume a relaxed position.

Repeat these four phases several times, slowly and consciously. You can, of course, repeat phase two and three a few times before concluding the Routine with phase four.

This Mensendieck Routine aims at strengthening the muscles between your toes; imagine them in the balls of your feet.

NOTE: It is not unusual that beginners are plagued by cramps in their feet muscles while executing these exact foot movements. This usually means that your foot muscles are in such poor condition that they can't cope with your demands. The gentler you proceed with calling your muscles to task, the more willingly they will respond. In the course of weeks, you can gradually increase the tension in your muscles while executing the Mensendieck Techniques, continually staying under the cramp level. The healthier your muscles become, the more the cramp level will retreat.

Mensendieck Technique 11: ANKLE CONDITIONERS
A. Supination and Pronation
B. Plantar Flexion and Dorsiflexion

PREPARATORY POSITION: Start in the same position as required for Technique 10. Then, place your feet into the Well-balanced Foot Position, as instructed in Technique 9. Throughout both Routines your feet must maintain their cups by keeping the toes extended in line with the instep.

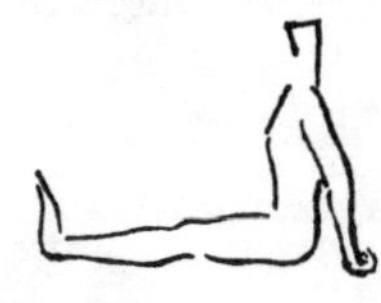

A.

• Slowly contract the muscles along the inside of your ankles, thereby pulling the inner margins of the feet closer to their inner ankle, i.e., supination, while maintaining your cupped feet at right angle to leg position.

• Slowly contract the muscles along the outside of your ankles, thereby pulling the outer margins of the feet closer to their outer ankles, i.e., pronation, while maintaining your cupped feet at right angle to leg position.

Repeat these two phases a few more times before finishing.

This Mensendieck Routine aims at strengthening the muscles on either side of your ankles, the stirrup muscles. Once the Technique is mastered, try to apply resistance: when the muscles along the inside of your ankles perform the motion, resist this motion with the muscles along the outside of your ankles and vice versa.

B.

• Slowly contract the muscles in your calves, thereby pulling your heels straight backward, closer to the lower legs and extending the cupped feet downward, in line with the legs, i.e., plantar flexion.

• Slowly contract the muscles along the front of your ankles, thereby raising your cupped feet back into the Well-balanced Foot Position, i.e., dorsiflexion.

Repeat these two phases several times, preferably while applying resistance: when the muscles along the back of your ankles perform the motion, resist this motion with the muscles along the front of your ankles and vice versa.

This Mensendieck Routine aims at strengthening the muscles along the back and the front of the ankles. Most people activate the muscles along the front of the ankles far too little.

Mensendieck Technique 12: THE PAN SCRAPE

PREPARATORY POSITION: Sit on the floor, with straight legs and back; feet in the Well-balanced Foot Position. Then cross one leg over the shin of the other leg and practice with the foot of your top leg (which is still straight and resting on the bottom leg). Keep your toes in line with instep throughout the whole Routine.

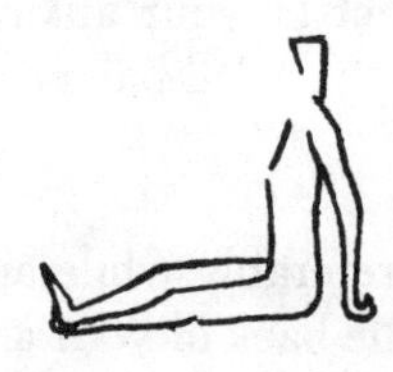

• Slowly contract the muscles along the inside of your ankle, thereby pulling the inner margin of the foot closer to its inner ankle.

• Slowly contract the muscles in your calf, thereby pulling the heel closer to the lower leg, while holding your firm grip on its inner margin.

• Slowly contract the muscles along the outside of your ankle, thereby pulling the outer margin of the foot closer to its outer ankle.

• Slowly raise the foot all the way in its ankle joint, while holding your firm grip on its outer margin.

• Carefully return the foot to the Well-balanced Foot Position, by somewhat lessening your firm grip on its outer margin.

Repeat these five phases, slowly and consciously. Then practice with the same foot the other way around, before crossing your legs the other way and giving your second foot a similar workout.

The moving foot should describe a rectangle rather than a circle; imagine that you are scraping out a pan with your foot.

This Mensendieck Routine aims at strengthening all the muscles in your lower legs and feet. Its effectiveness is increased if you apply resistance in and around your ankles with the respective opposing muscle groups. Practice the Pan Scrape as often as the chance presents itself, for instance, while watching television. You don't necessarily have to sit on the floor. Sitting cross-legged as in Technique 8 will do, and it provides you with more frequent opportunities to practice.

Mensendieck Technique 13: CATERPILLAR

PREPARATORY POSITION: Sit well balanced on a stool, feet parallel on the floor, with your heels directly below the knees. Realize that each foot has two arches, to wit: a longitudinal arch between the heel and the base of the big toe, and a transverse arch between the base of the little toe and the base of the big toe.

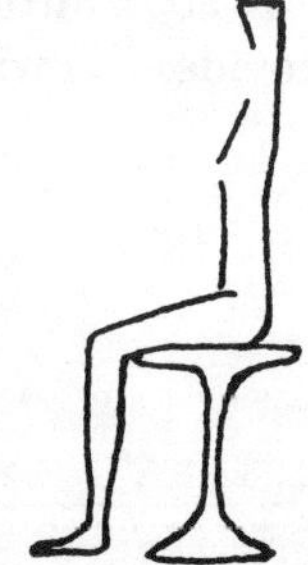

• Concentrate on the sole of one of your feet and become conscious of the in-touch-with-floor sensation in your heel and in the base of your big toe; be aware of the span between these two points of supports.

• Raise this span by contracting the muscles in the longitudinal arch of the foot and allowing your heel to slide over the floor, straight forward toward the base of your big toe while keeping your toes straight. Automatically, the unloaded outer margin of the foot will have approached the base of the big toe, thereby causing or enhancing the transverse arch of your foot.

• Maintain the tension in both arches of the foot for a few seconds.

• Allow the forefoot to slide straight forward by releasing the tension in the arches of the foot.

Repeat this fourfold procedure about seven more times, slowly and consciously, and remember that the toes must remain extended in line with the instep throughout the Routine. Then return the foot to its preparatory position, and "caterpillar" similarly with your other foot.

This Mensendieck Routine aims at strengthening the muscles in the arches of your feet. A foot is, ideally, a rather pliant bridge which can smoothly absorb the impact of being loaded by sagging slightly in its longitudinal as well as its transverse arch. The muscles in these arches have to function as "springs" and return the bridge to its earlier shape when the foot is being unloaded. The Caterpillar is an excellent Technique to gain the necessary tone, thus resiliency, in these springs.

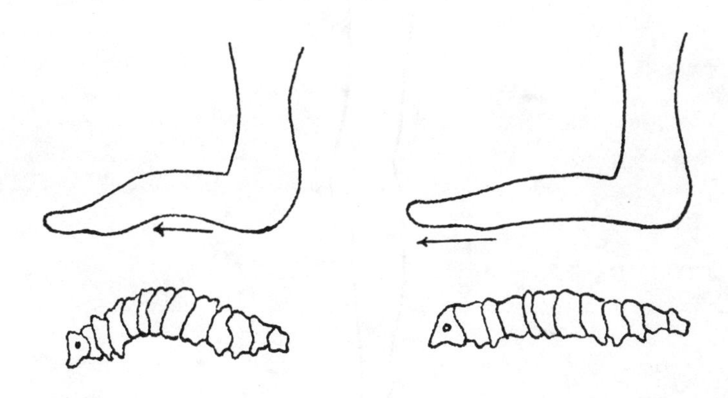

NOTE: Many people, through years of neglect, have allowed their feet to lose their ideal shape and capacity for pliancy. If your "foot bridges" are already rigid, your Caterpillar effort may not result in a visible raise of the arches with the corresponding forward slide of your heels. If so, you will have to start with cultivating muscle awareness in the spring muscles, through "listening to what it feels like" during your conscious build-up and subsequent release of the tension, while imagining the caterpillarlike movement. Practicing this Routine daily, with perseverance, will ultimately give visible results.

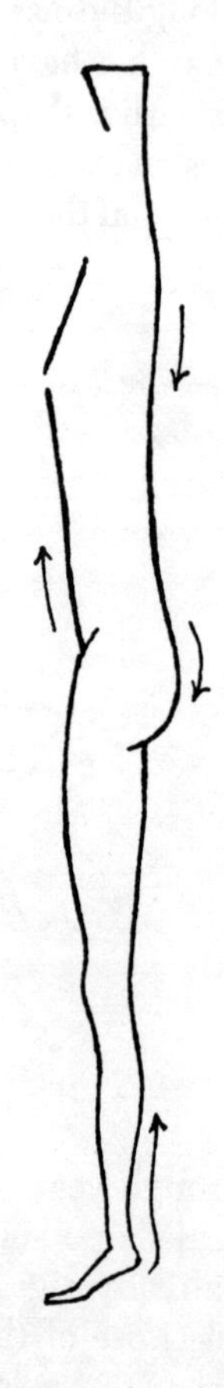

A well-aligned body can be raised in the ankle joints without any prior change in posture or shift of weight.

Mensendieck Technique 14: STANDING ERECT ON THE BASES OF THE BIG TOES

PREPARATORY POSITION: The Well-balanced Stance. (Please review Techniques 1 and 8B.)

• Inhale, while concentrating on the firm contact between the bases of your big toes and the floor.

This bases-of-big-toes-to-floor contact is always of utmost importance for maintaining your equilibrium, particularly while executing this Technique. It helps to prevent "wobbling" in your ankles.

• Exhale, while raising your heels with the calf muscles and those along the outside of your lower legs (peroneus muscles) until you are standing on the bases of your big toes, i.e., with the heels about 1½ inches off the floor.

Allow your whole body to be lifted in the ankles, which will happen if you maintain your well-balanced posture tall to sternum as well as to crown of head. In other words, allow your crown to approach the ceiling when you raise your heels.

• Inhale, while remaining stationary on the bases of your big toes.

• Exhale, while slowly lowering your body in the ankle joints; aim at keeping your feet perfectly parallel and at gently landing your heels only, so that the bases of your little toes remain just off the floor, assuring proper peroneus action.

The same muscles which raised your body in the ankle joints are engaged for the controlled lowering of your heels, mainly by resisting gravity.

Major calf muscles
(gastrocnemius muscle and soleus muscle—underneath)

If you succeeded in keeping your feet parallel, the heels are now again obscured from view in your front mirror image.

This Mensendieck Routine challenges all your muscles to sustain posture, while the demand for a perfect balance is intensified. The Technique is ideal to check whether or not your Well-balanced Stance is correct, because, when your body masses are properly superposed and your weight is divided equally over your four points of support (heels and bases of big toes), you can lift yourself in the ankle joints without any prior change in posture or shift of weight. Thus make it a point to often lift yourself correctly onto the bases of your big toes, to check your alignment and balance. And each time while lowering the heels straight backward, be consciously aware that the bases of your little toes remain all but unloaded.

Never raise your heels higher than about two inches, when your body is standing erect. Any higher raising will lift you onto your toes, which cultivates spread-feet or the loss of the transverse arches in the balls of feet. These are the arches which will gain in quality and shape when we cup the feet often.

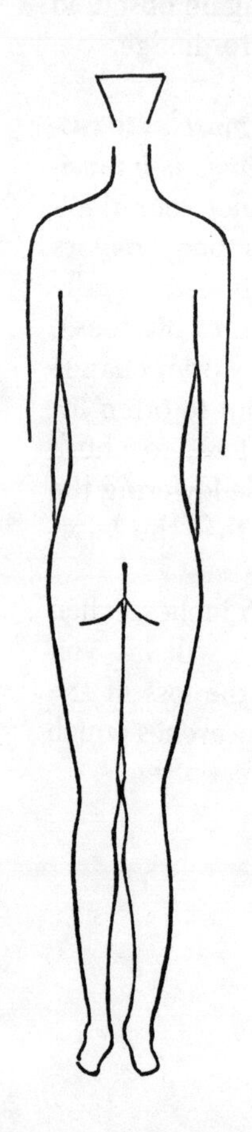

Release the tension in your corset muscles, allowing your knees, inner ankles, and foot main arches to sag inward and down.

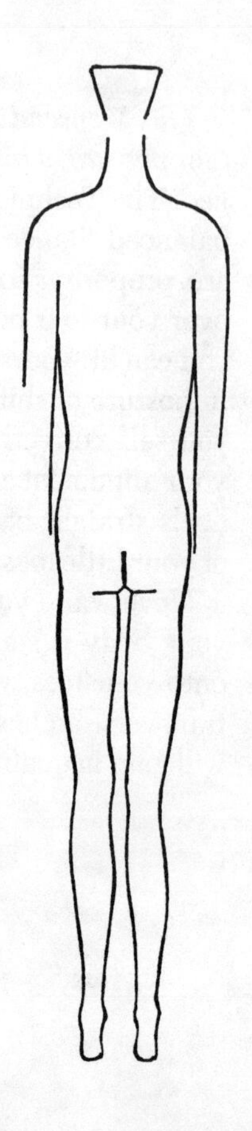

Contract your corset muscles and lift your knees, inner ankles, and foot arches in their proper positions while taking care not to load the bases of your little toes.

Mensendieck Technique 15: ANKLE CARRIAGE

PREPARATORY POSITION: The Well-balanced Stance. Observe the position of your inner ankles in the front mirror image.

• Inhale, while slowly releasing the tension in your corset muscles (abdomen, buttocks), thereby allowing your knees to turn slightly inward and your inner ankles to sag toward each other and down. This will lower the longitudinal arches of your feet. Your inner ankles now feel more heavily loaded than before.

• Exhale, while slowly drawing your abdominal muscles upward and your buttocks firmly together and slightly under, as if you are wrapping them around your hips. This will twist your legs forward and outward in their articulation with the pelvis. Allow this until the kneecaps face straight forward again, and the inner ankles and foot arches are lifted back in their proper positions.

No inward sliding of heels, or weight shift toward outer margins of feet, or lifting of bases of big toes may occur during this "spiraling your inner ankles upward" phase. Thus, loading the bases of your little toes is incorrect! Your four supports are: two heels and two bases of big toes.

Repeat these two phases several times. You may bend the knees a little during the release phase, to gain a more distinct sensation of heavily loading your feet as soon as your corset muscles quit carrying your inner ankles and your knees. Then straighten the knees till normal during the tension phase.

This Mensendieck Routine teaches you to carry the arches of your feet high, **mainly** *with your corset muscles around the pelvis which, by the way, will be conducive to keeping your hips trim.* Practice this Technique routinely whenever you stand: while ironing, or waiting for the milk to boil or for the pedestrian light to turn green, or while attending a cocktail party (in low heels!). But be sure that your feet remain parallel; the movement cannot be properly effectuated if your feet are toeing out.

At this point in your progress, let's review the correct parallelness of your feet once more, and more specifically this time: Parallel position of the feet is the placement of the feet with toes pointing straight ahead and heels pointing straight backward. When you stand correctly, the balls of the feet will usually have to be almost two inches apart while the distance between the heels exceeds two inches. The requirement is to have your knees and ankle joints as well as the longitudinal axis (from middle of heel to second toe, adjacent to big toe) of your feet straight below your hip joints. Your hip joint on either side is approximately at midpoint between your pubic symphysis and the outline of your "hips," thus far more inward than most people realize.

Not only the face, but the entire body, can be animated to expression by the soul.

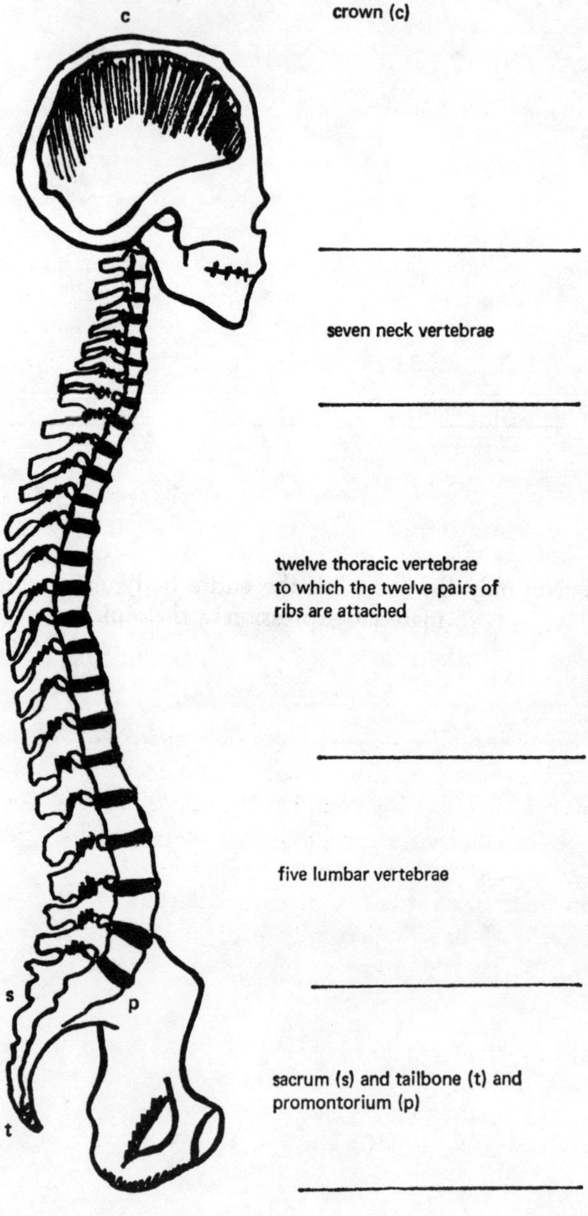

c

crown (c)

seven neck vertebrae

twelve thoracic vertebrae
to which the twelve pairs of
ribs are attached

five lumbar vertebrae

s

p

sacrum (s) and tailbone (t) and
promontorium (p)

t

XVIII. YOUR SPINAL COLUMN AND ITS DISKS

The spinal column encloses the neural canal, which contains the spinal cord. It is composed of a series of bony masses called vertebrae. The bottom portion of the spinal column, known as the sacrum, is made up of five fused vertebrae and has the shape of a curved triangle; it is wedged dorsally between the two hipbones. Its apex, the coccyx or tailbone, consists of a few rudimentary tail vertebrae, which are fused to each other and to the sacrum. The center of the body is considered to be at the promontorium, which is the prominent anterior border of the body of the top sacral vertebra. Above the sacrum are twenty-four true vertebrae, which become increasingly smaller in size. On top of these rests the head, which for all practical Mensendieck purposes must be considered to be your uppermost vertebra, because this image is of great help in correctly aligning your spinal column from the tailbone to the crown of your head.

In between adjacent vertebrae are fibrocartilaginous pads called intervertebral disks, each of which consists of a fibrous ring enclosing a pulpy center. These disks serve as shock absorbers and make intervertebral movement possible. However, the very fact that they are easily subject to distortion makes them quite vulnerable. Constant distortion of one or more disks, due to habitually faulty superpositioning of the vertebrae, can have ruinous consequences.

Ideally, the weight of higher body masses should always be equally distributed over the entire area of each intervertebral disk. This is the main reason why it is of crucial importance always to align your vertebrae correctly when your trunk is

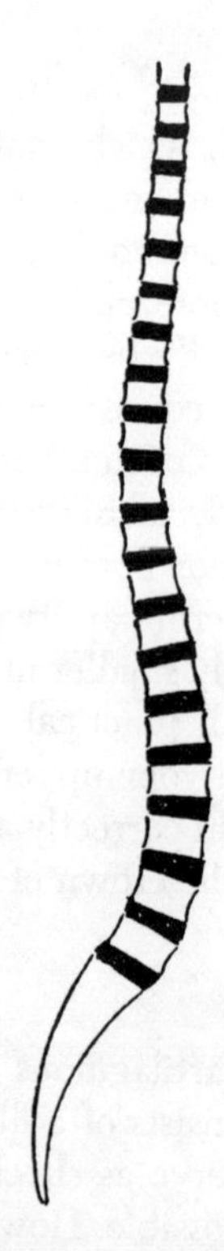

when the body is upright and correctly aligned, the angle that the top surface of the sacrum makes with the horizontal is approximately 30 degrees, allowing a smooth compensatory curving of the vertebral column and thus a rather equal weight distribution of the higher up body masses over all of the intervertebral disks

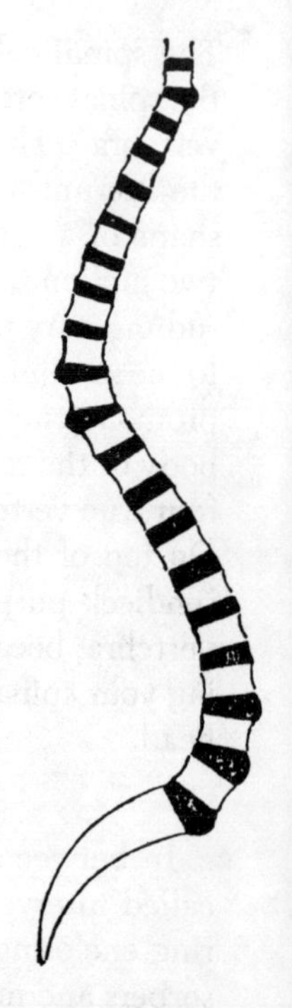

here you can see how extreme compensatory curving of the vertebral column, necessitated by a too slanted sacrum position, makes it impossible to distribute equally the weight of the higher up body masses over hardly any of the intervertebral disks

Habitually aligning your vertebral column correctly, which is dependent upon a perfect relationship of your legs with your feet and of your sacrum (thus pelvis) with your legs, will help prevent herniated disks.

stationary and upright and always to realign your vertebrae correctly following vertebral motion. Also this is why the vertebral column must be maintained tall to crown, whether stationary or in motion, to prevent needless compression of the intervertebral disks.

If, through faulty posture, the weight of higher body masses is **not** equally distributed over a disk, one part of the disk carries a very light load, whereas the remainder of the disk is carrying far more weight than it may be able to withstand in the long run. This results in gradual degeneration and weakening of the disk as well as of the surrounding ligaments. Eventually, the fibrous ring of the disk may split, and the pulpy nucleus will then be extruded through the ligaments. Usually the extrusion progresses gradually, causing chronic back pain in and around the affected area. When the extrusion happens posteriorly, disk material will protrude into the neural canal, and the resulting pressure on one or more nerve roots causes the true disk syndrome, with pain and other symptoms along the course of the affected nerve(s).

The disk between the sacrum and the lowest lumbar vertebra and the disk between the two lowest lumbar vertebrae are subject to tremendous forces and are consequently particularly prone to degenerative changes. It is here that the vertebral column starts its first countercurve to compensate for the oblique position of the curved sacrum. The sacrum cannot be modified, but its position depends upon how the pelvis is carried relative to the legs. Technique 16 will teach you the difference between a forward tilted and an upright pelvis position. Make it a point from now on to tilt your pelvis only when specifically required in a Mensendieck Technique for the purpose of maintaining normal mobility and adequate muscle resiliency in the lower back. Throughout daily life activities you must carry your pelvis properly upright with your corset muscles, so that the weight of higher body masses can be distributed over the entire area of these two most vulnerable disks. Once the foundation of your spinal column is properly carried (legs relative to feet, pelvis relative to legs), this column can assume correct alignment from tail to crown, essential for the proper loading of all your disks.

Chronic back pains resulting from early disk degeneration usually respond favorably to Mensendieck therapy. The true disk syndrome, where nerve roots are already impinged upon, can seldom be reversed by any conservative treatment; however, corrected postural alignment and improved skill in functional body movement usually result in a reduction of the pain and secondary symptoms induced by the herniated disk, as well as those induced by your efforts to escape the pain.

When surgical treatment has been performed to eliminate the cause for the true disk syndrome, Mensendieck will help you to regain normal mobility and to prevent recurrence of the syndrome.

The minds of all men should be aroused to understanding muscles as the controlling factor for the human upright structure.

ON HANDS AND KNEES

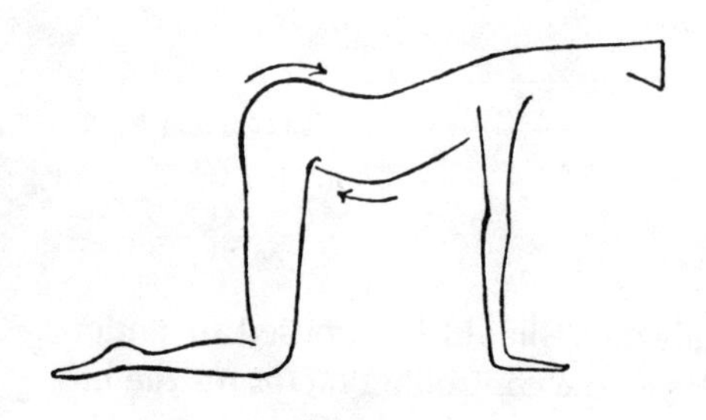

inhale and tilt pelvis forward

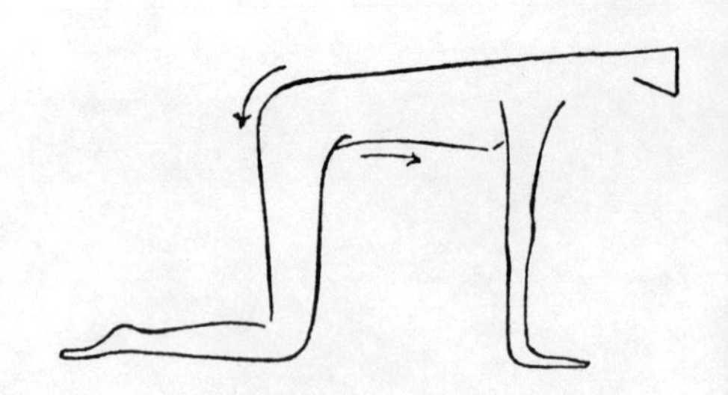

exhale and raise pelvis until preparatory position is resumed

Mensendieck Technique 16: PELVIS ROCKING
A. On Knees and Hands
B. Supine
C. Standing

A.

PREPARATORY POSITION: On knees and hands, with your upper legs and arms vertically under your hip joints and shoulder joints respectively, feet pointing straight backward and hands pointing straight forward; carry your neck and head in line with the straight spine.

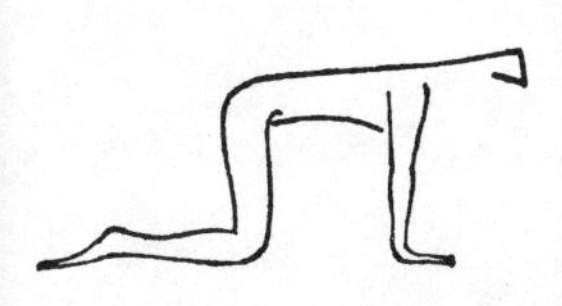

• Inhale, while slowly releasing the tension in your buttocks and abdominal muscles, thereby allowing your pelvis to tilt forward in the hip joints, so that the shape of your lower back changes from straight to saddle-shaped or concave.

If you experience difficulty in finding the movement, think of your tailbone and point it toward the ceiling.

• Exhale, while slowly drawing your buttocks under and your abdominal wall tight, until your spine is straight again, from tail to crown.

Repeat these two phases several times, maintaining your body weight equally distributed over the knees and hands, and your neck tall to crown, shoulders low.

SUPINE

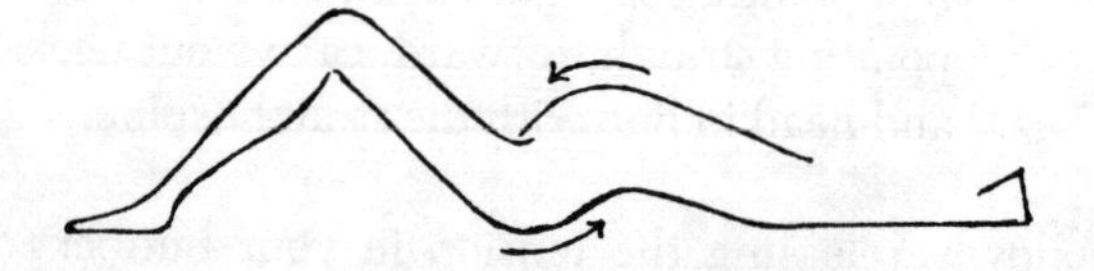

inhale and tilt pelvis forward

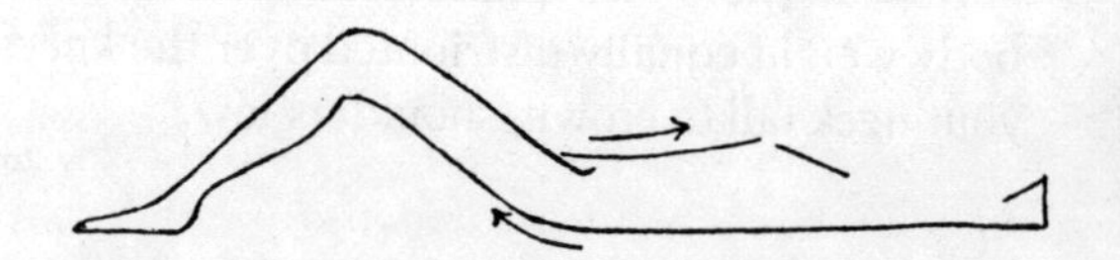

exhale and raise pelvis until preparatory position is resumed

B.

PREPARATORY POSITION: Supine. Draw your knees up, one at a time, while exhaling and contracting the abdominal muscles, until the soles of your feet and your lumbar vertebrae are on the floor. Arms alongside your body, palms down on the floor. Shoulder blades low and next to your spine. Neck tall to crown.

• Inhale, while slowly releasing the tension in your buttocks and abdominal muscles, thereby allowing your pelvis to tilt forward in the hip joints, so that your lumbar vertebrae leave the floor to shape the lower back concave.

Rather than forcing your lower ribs in the direction of the ceiling, think of moving your pubic bone toward your feet, while allowing your chin to approach your sternum a little.

• Exhale, while drawing your buttocks under and your abdominal muscles toward your sternum until the lumbar vertebrae are again in contact with the floor.

Exhale completely at the conclusion of this phase, so that your belly muscles get a firm grip on your lower ribs.

Repeat these two phases several times, slowly and consciously. Become aware that the primary movement of your pelvis is in the hip joints; it helps to think of an axis connecting these two joints and to rock your pelvis forward and backward on this axis.

STANDING

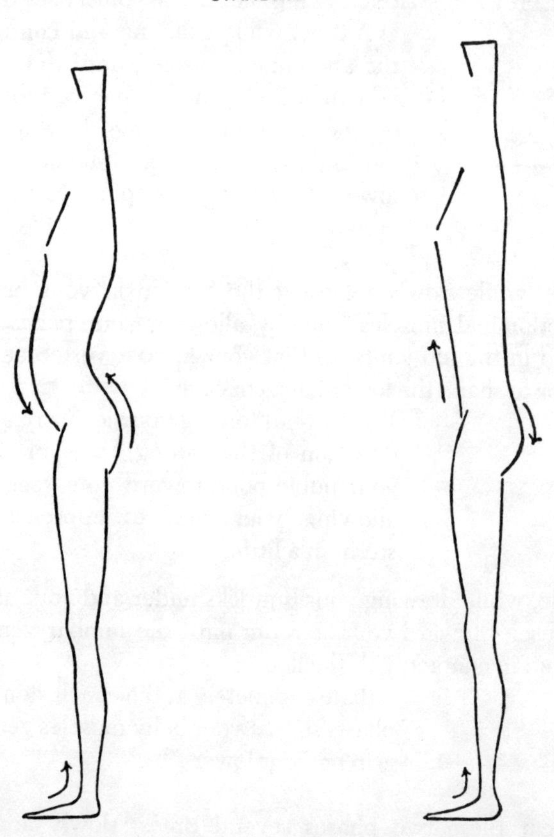

inhale and tilt pelvis forward

exhale and raise pelvis until preparatory position is resumed

The Mensendieck Routines 16A and 16B aim at releasing accumulated strain in the muscles of the lower back and at acquainting you with proper abdominal contraction. Once you have familiarized yourself with how the pelvis can rock in the hip joints, and how this causes a secondary movement in your lumbars, try the Pelvis Rocking while standing:

C.

PREPARATORY POSITION: The Well-balanced Stance.

• Inhale, while slowly releasing the tension in your buttocks and abdominal muscles, thereby allowing your pelvis to tilt forward in the hip joints, so that your lower back becomes more concave (like a sway-back).

It helps to think of moving your tailbone in the direction of the ceiling, rather than backward. Allow no movement in your ankle joints, so that your knees will not hyperextend and your body weight will remain equally distributed over heels and balls of feet.

• Exhale, while drawing your buttocks together and under, and your abdominal muscles from groin to chest, thereby raising the pelvis in the hip joints to its normal upright position with your lower back only shallowly concave.

Repeat these two phases several times, keeping your legs and the upper part of your trunk exactly in the position assumed in the Well-balanced Stance. It is advisable to try out the Pelvis Rocking while standing close to the back rest of a straight chair. After constructing the Well-balanced Stance, rest your hands on the back of the chair. With the exception of your pelvis and lumbars, all your other body masses must remain in the same relationship to the chair when practicing the two phases of the Technique. Once you succeed in this, extend your arms alongside your body, and again practice the Pelvis Rocking without moving in your ankles and upper trunk. Eventually you will no longer need the orientation relative to the chair.

Mensendieck Routine 16C aims at releasing accumulated strain in the muscles of the lower back and serves to make you more aware of how the pelvis normally should and should not be carried. Habitually carrying the pelvis in raised, i.e., upright, position as learned in phase two, will ease the posture-sustaining task of your lower-back muscles and will help to prevent intervertebral disk problems.

Postural beauty is attainable for all men.

IN TAILOR SIT

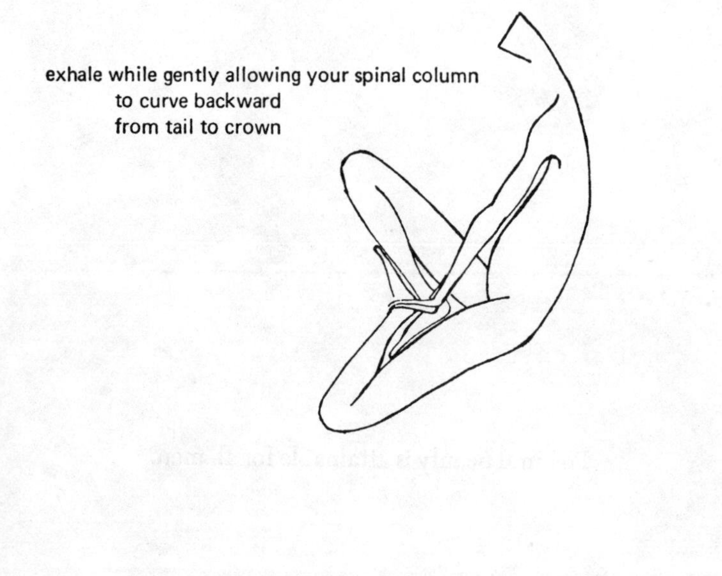

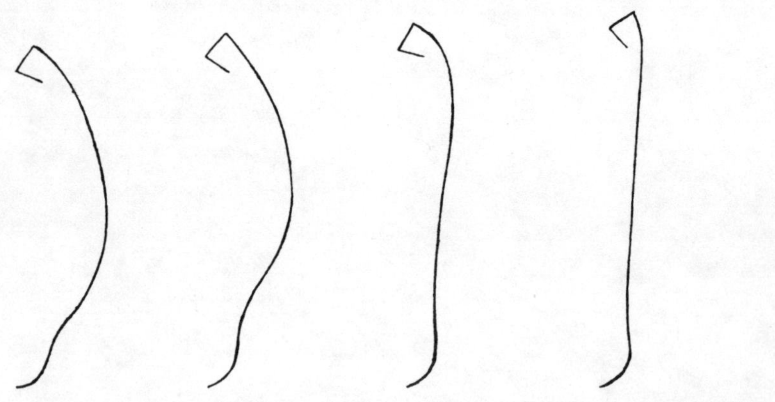

exhale while gradually straightening your spinal column from tail to crown (in vertical)

XIX. GAINING AWARENESS OF YOUR SPINAL COLUMN

Mensendieck Technique 17: STACKING THE VERTEBRAE

A. In Tailor Sit
B. In Zigzag Sit

A.

PREPARATORY POSITION: Sit cross-legged on the floor, with the sole of one foot against (not under) an inner thigh, other sole against a shin, knees low. Place one hand on the other, then rest both on your front ankle if that is easy, otherwise on the rear ankle. (They should be exactly in front of you.) Your back must be straight and vertical, your shoulders low.

• Inhale, and direct your thoughts to your spinal column, on either side of which run long and slender muscles (sacrospinalis group) from your pelvis to the back of your head. The beginning of your spinal column is at the tailbone, and its top for our purposes is at the crown of your head.

• Exhale, while gently allowing your spinal column to curve backward.

Initiate this movement by rolling off your sedentary bones, which will result in a backward sagging of the sacrum and the lumbar vertebrae as well as the lower thoracic vertebrae. Then bend the upper

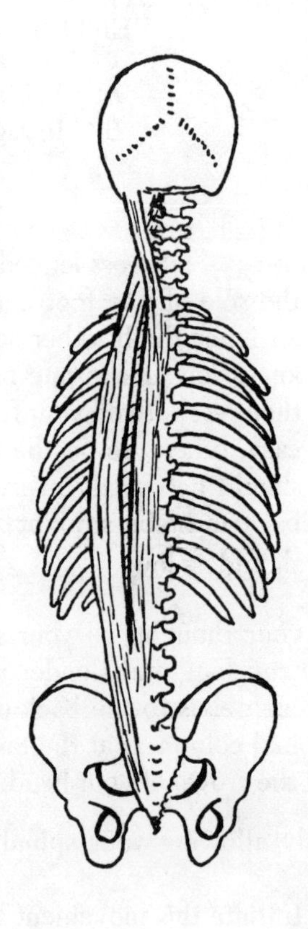

Left sacrospinalis or erector spinae muscles

thoracic vertebrae and your neck and head forward. The latter should remain above the pelvis throughout this Routine.

• Inhale, while your back remains relaxed and curved, shoulders low, arms relaxed.

• Exhale, while gradually straightening your spinal column from tail to crown to resume the Preparatory Position.

Initiate this movement by lifting the sacrum out of its backward tilt to upright, which will provide you with the proper base on which to stack the five lumbar vertebrae, on these the twelve thoracic vertebrae, then the seven neck vertebrae and finally your head. Your shoulders should remain low, your arms relaxed, your head above the pelvis.

Repeat these four phases a number of times, and remember to move both ways from tail to crown—vertebra for vertebra—while exhaling.

This Mensendieck Routine teaches you to move the segments of the spinal column relative to one another and aims at proper usage of the erector muscles of your back. If you can't sit comfortably on the floor in the position described, start out from and return to sitting well balanced on a stool (heels below knees), one hand resting on your thighs near each knee. Execute the Routine in this position whenever you can spare a minute, while sitting behind your desk or sewing machine for instance. It is an ideal movement to release accumulated tension in back muscles due to sitting in one position for an extended period. After having mastered Technique 17A, whether seated on the floor, tailor-fashion, or on a stool, practice the same in a different position, the zigzag sit:

IN ZIGZAG SIT

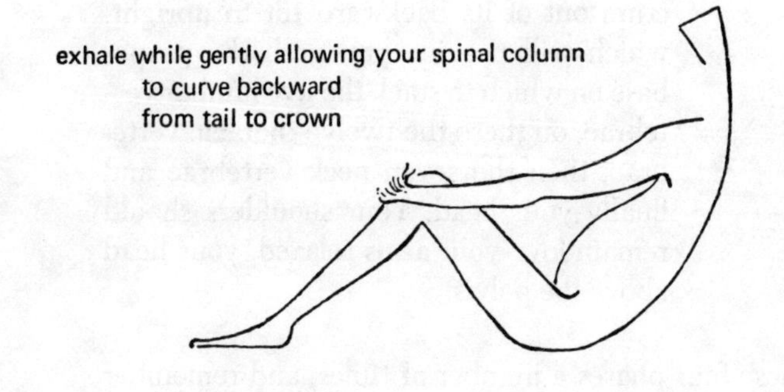

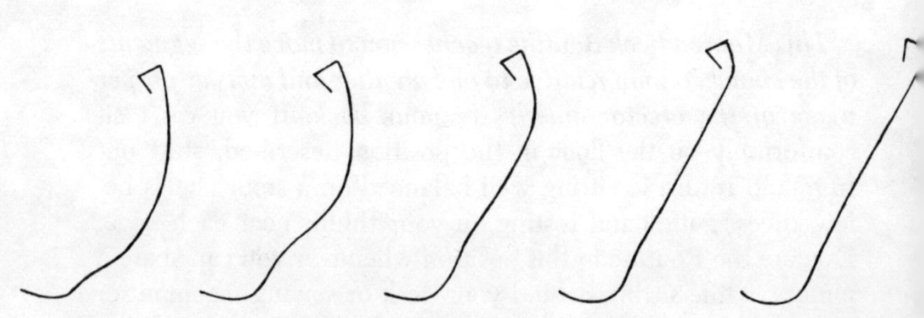

B.

PREPARATORY POSITION: Sit on the floor. Draw your knees up until the soles of your feet rest on the floor. Wrap your hands (palms) around and not below the knees, with the fingers interlocked. Lean backward until your elbows are straight. Shoulders low, throughout the Routine.

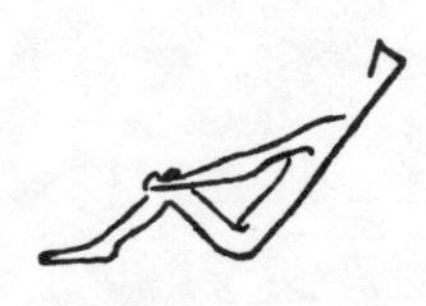

• Inhale, and direct your thoughts to your spinal column.

• Exhale, while gently allowing your spinal column to curve backward, tail to crown.

• Inhale, while your back remains relaxed and curved.

• Exhale, while gradually straightening your spinal column, tail to crown.

Keep your arms relaxed throughout the Routine, thus maintaining distance between knees and sternum. When straight, your back will be oblique rather than at a right angle to the floor (as was the case in 17A).

Repeat these four phases a number of times, moving both ways from tail to crown while exhaling.

This Mensendieck Routine aims at strengthening the erector muscles of your back. Eventually, these muscles should be capable of maintaining your spine absolutely straight when you take your hands off your knees for a moment.

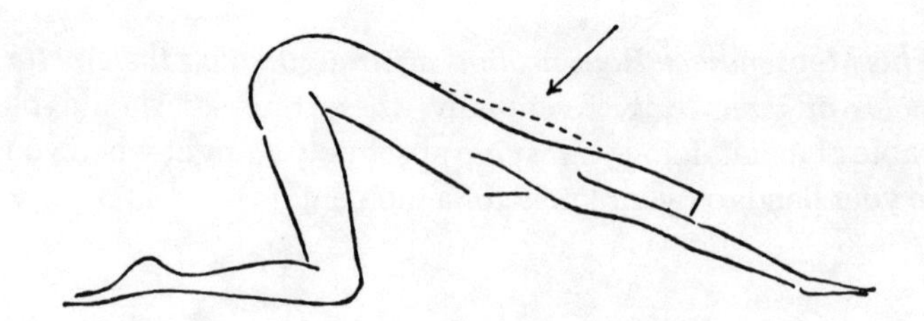

Bounce your thoracic vertebrae toward your knees.

XX. CORRECTING THE RIGID UPPER BACK

During the practice of Technique 17, you may have discovered that the thoracic section of your spine resisted your effort to straighten it up. Usually the cause for this rigidity is that you have allowed yourself to habitually stand and sit stoop shouldered as well as hunched back, which most probably also has brought about a shortening of the muscles along the front of your shoulders (pectoral muscles). This can be corrected by carrying your body properly aligned throughout your daily activities. In addition, frequent practicing of the following two Routines can help you to regain normal resiliency in the upper back and in the pectoral muscles.

Mensendieck Technique 18: THE CHEST BOUNCE

PREPARATORY POSITION: On the knees. Place the hands way out in front of you on the floor so that your trunk and arms are spanned from hips to wrists: Distributing your weight equally between knees and hands requires the upper legs to deviate slightly backward from the vertical. Carry the neck and head-to-crown in line with your spine.

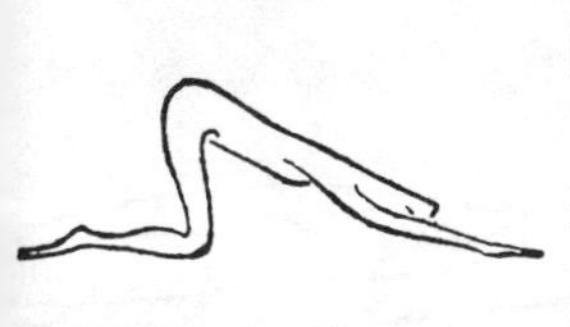

• Concentrate on your thoracic vertebrae.

• Bounce your thoracic vertebrae, in a rhythmic sequence, downward (only!) in the direction of your knees.

Continue this rhythmic downward bouncing for a while before returning your trunk to its normal upright position, in line with the upper legs. Then prepare for Technique 19 by lowering your body to the floor till you lie supine (as instructed in chapter XII).

Mensendieck Routine 18 will particularly limber up the thoracic section of the spinal column in addition to stretching the pectoral muscles.

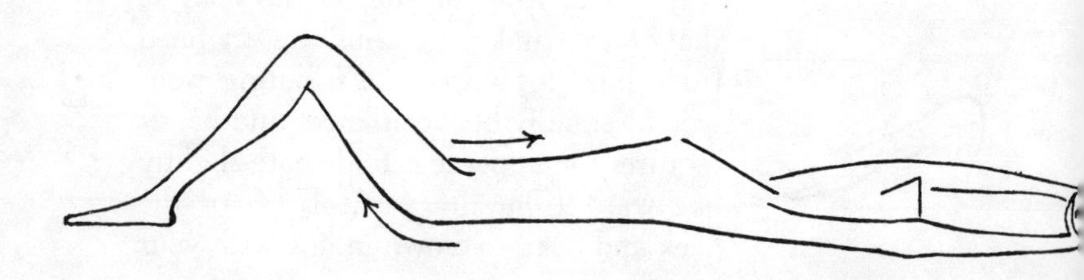

Slide your hands away from your head for as long as your lumbar vertebrae and your elbows can remain in contact with the floor.

Mensendieck Technique 19: STRETCHING THE PECTORALS

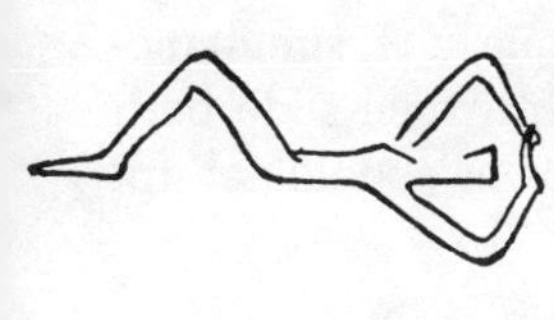

PREPARATORY POSITION: Supine. Draw your knees up, one at a time, while exhaling and contracting the abdominal muscles, until the soles of your feet and your lumbar vertebrae are on the floor. Interlock your fingers, just above the crown of your head, and turn your linked hands over. Ideally, your little fingers and elbows are now on the floor. Inhale.

• Exhale, while contracting your abdominal muscles and firmly drawing the pubic bones toward your sternum, to secure the pelvis and to stabilize the lumbar vertebrae against the floor; then slide your hands away from your head, little fingers over floor. Proceed for as long as your elbows can remain in contact with the floor. Try to go as far as you possibly can, gently though decisively.

If the muscles in the front of your armpits are *not* in particular need of regaining normal length, you will be able to extend your arms all the way, with the little fingers and elbows as well as your lumbar vertebrae (!) in touch with the floor.

• Inhale, while returning the back of your linked hands to the crown of your head.

Those who become aware of their particular need for this Routine should not repeat these two phases too often in the beginning. The movement is more powerful than you may realize initially.

Mensendieck Routine 19 particularly aims at regaining normal length in the pectoral muscles which run from the chest toward the upper arms. Fortunately, muscles which were allowed to become shorter than normal can be elongated, if you regularly invite them to do just that.

Before you get up from the floor, as instructed in chapter XII, the legs first have to be straightened in the knees and hips. This can be executed by giving your feet arches and belly muscles a workout as follows: Caterpillar (Technique 13) simultaneously with both feet until no longer possible, then slide both heels over the floor while emphasizing a lower abdominal upward pull and exhaling.

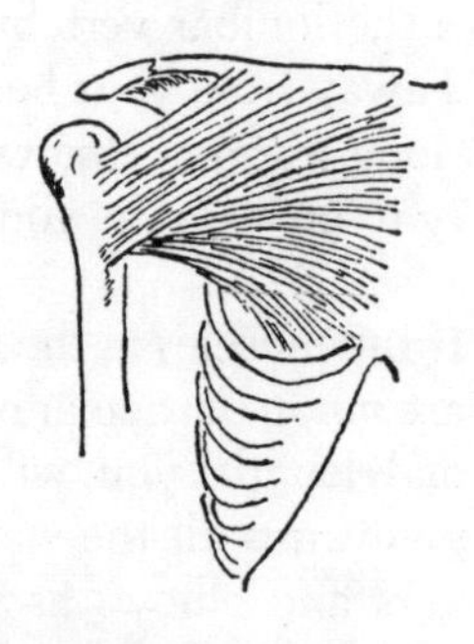

Right pectoralis major

XXI. RELIEVING BACKACHES OF MUSCULAR ORIGIN

Generally, the reaction to a sensation of discomfort is to hold the body rigid, and a rigid body cultivates muscular pains. As a rule, if you sense discomfort in your muscles, move. Usually, this will bring about the longed-for relief immediately. If you are already in pain (although at first it can be agonizing to break through such a barrier), move time and again in and around the affected area until you have eventually released the muscular pain. Movement cleanses and nourishes your muscles back to a painless and healthy condition. (See chapter III.) Any off-and-on muscular activity is better than no movement at all. Selecting the appropriate Mensendieck movements will help you get rid of these wholly unnecessary muscular pains faster.

If you are in doubt as to whether your pain is caused by a poor muscular condition, consult your physician.

Since most muscular aches are located in the back, we will discuss here some simple Mensendieck Routines for that area in particular. These, as well as Techniques 6, 7, 16, 17, 18, and 19, are excellent to practice regularly if you want to prevent getting backaches, or prevent their recurrence. If you feel a strain in your muscles while practicing a particular movement, those muscles are getting the action they need most.

THE DIRECT CHANGE-OVER

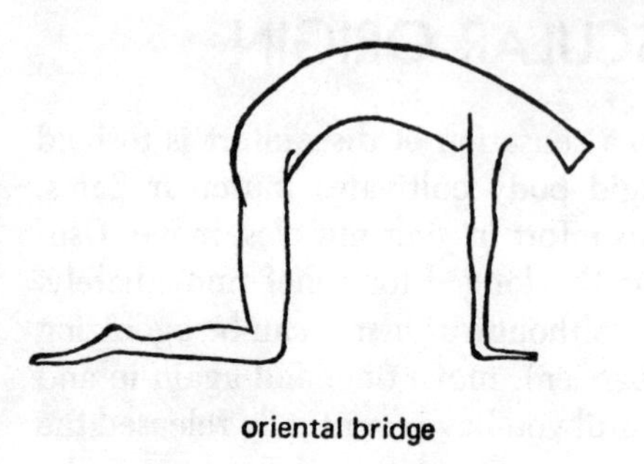

oriental bridge

suspension bridge

Mensendieck Technique 20: THE BRIDGES
A. The Direct Change-over
B. The Indirect Change-over

PREPARATORY POSITION: On knees and hands, with your upper legs and arms vertically under the hip joints and shoulder joints respectively, feet pointing straight backward and hands pointing straight forward. Your tall neck and head in line with the straight spine.

A.

• Exhale, while slowly drawing your tailbone diagonally downward in one direction while at the other end your neck and head unit are extended diagonally downward in the other direction. This will buck the lumbar vertebrae (in particular) firmly upward, i.e., convex. Your back is now shaped like an oriental bridge.

• Inhale, while slowly extending your tailbone as well as your neck-head diagonally upward, thereby allowing your thoracic vertebrae (in particular) to gently sag way down, i.e., concave. Your back is now shaped like a suspension bridge.

Repeat these two phases, slowly and conscientiously, until you have mastered the basic Technique. Then proceed with Technique 20B.

THE INDIRECT CHANGE-OVER

construct oriental bridge . . . and . . . bring tailbone close to heels, then

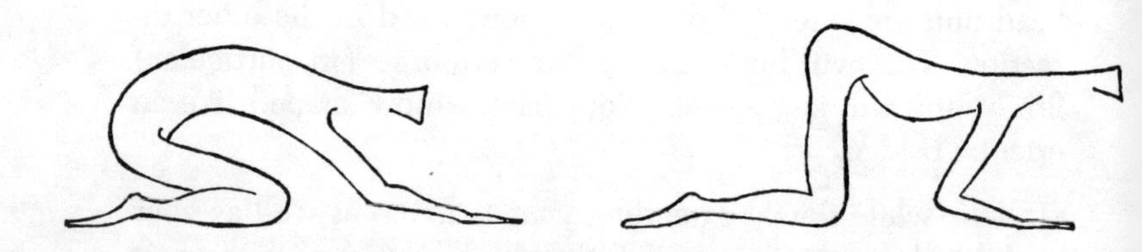

bend elbows sideways and change shape of your back while moving it forward;

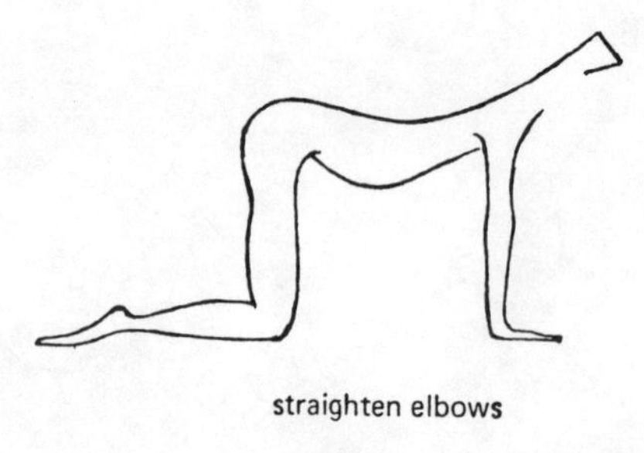

straighten elbows

B.

• Exhale, while first shaping your back into the "oriental bridge." Then bring the tailbone close to your heels, leaving your hands exactly where they were.

• Bend your elbows sideways without resting them on the floor.

• Inhale, while changing your back to concave, from tail to crown, during your trunk's forward movement, close to the floor, until your weight is again equally distributed over the knees and the hands.

• Straighten your elbows. At this point your back is in the suspension-bridge shape.

Check yourself: Are your thoracic vertebrae deepest down at the conclusion of phase four? Usually this can be improved upon by extending the tailbone and the neck-to-crown more diagonally upward. The head should not be raised in relation to the neck.

Repeat these four phases several times.

These Mensendieck Routines aim at limbering up your spine and at stretch-relaxing all the muscles in your back. Remember, you are as young as your spine is flexible. *Gentle though decisive elongation of muscles brings about a release of tension in many instances.*

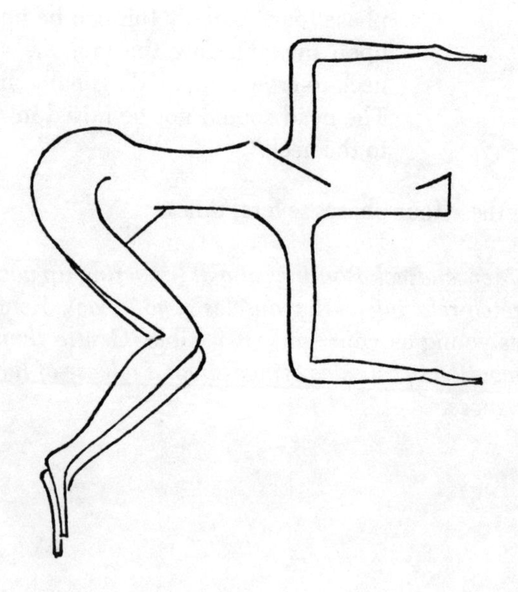

The supine stretch-twist position, assumed at the conclusion of phase two.

Mensendieck Technique 21: SUPINE STRETCH-TWIST

PREPARATORY POSITION: On your back. Knees bent, soles of feet on floor, neck tall. Your arms on the floor as follows: first extend the arms sideways on shoulder level, palms up; then bend your elbows, by moving the back of your hands upward over the floor, until the forearms are at right angle with the upper arm. Shoulders low, blades to spine.

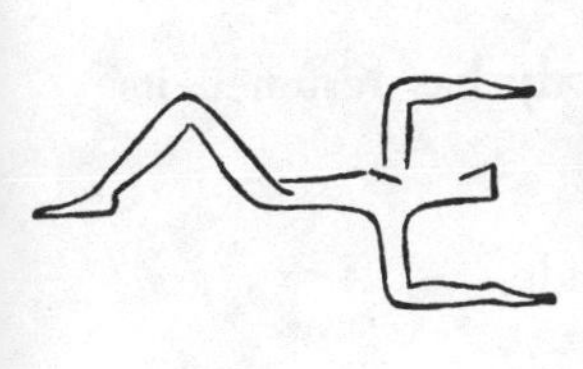

• Step around sideward with the feet, while keeping the knees bent, until you can step no farther. Your pelvis will have tilted sideways during this phase.

The side step is with both feet in the same circular direction. For instance, take small step to left with left foot, then place right foot next to the inner margin of the left foot. Repeat this about two more times.

• Gently lower both bent knees toward the floor, near your elbow, thereby allowing your spine to fully twist. Only one foot remains on the floor.

Ideally, the shoulder blade and arm on the opposite side are still on the floor.

• Lie in this stretch-twist for a while; allow the stretchings to happen.

If the opposite shoulder and arm come off the floor during phase two, relax and become aware of the sensation of heaviness in that arm; allow it to lower the shoulder and the arm. If you practice this Routine regularly, the shoulder and arm will eventually remain on the floor while you stretch-twist your back. You should also practice Routines 18 and 19 often.

• Lift your knees off the floor to get the soles of both feet on the floor.

• Step back with your feet until your body has resumed its preparatory position.

Repeat these five phases on your other side.

This Mensendieck Routine is a highly effective general "relaxant." Practice it after an exercise session or at the end of a strenuous day. *It can also be helpful to stretch-twist before getting up in the morning to dissolve possible stiffness in joints and muscles due to hours of inactivity.* The Technique can be easily practiced in bed. Two additional areas can be included in the Supine Stretch-Twist, at the end of phase two: while keeping your neck tall, turn your head far on its side, then on its other side. Straighten the top lower leg in line with its upper leg. Bend the knee again before proceeding with phase four.

Correct movements are the best means to preserve the body structure unimpaired.

Allow your pelvis to tilt with its side straight downward.

Mensendieck Technique 22: BUS STOP

PREPARATORY POSITION: The Well-balanced Stance. Carefully transfer your weight to the right foot only by allowing your upright pelvis to move a few inches sideways to the right. Raise your left lower leg straight backward in the knee joint until the ball and toes of your left foot can rest on the floor, on level with your right heel. You have now assumed the One-legged Stance (Mensendieck Technique 2).

• Inhale, while gently releasing the muscles of your right hip (minor glutei) and your left lumbar area (quadratus lumborum) and left inner thigh (adductors), thereby allowing your pelvis to tilt with its left side straight downward, which will bend the relaxed left knee a little more.

The left heel must maintain its original distance from the floor throughout the Routine; the left leg remains unloaded, with its foot parallel to your right foot.

• Exhale, while concentrating on contracting the minor glutei of your right hip, thereby allowing the pelvis and the left knee to slowly return to their respective positions, prior to phase one.

The left quadratus lumborum and adductors are also active during your return to the Preparatory Position; the better you master the Technique, the more you will learn to feel these muscles too. The left leg must not push that side of the pelvis upward.

gluteus medius
covering gluteus minimus

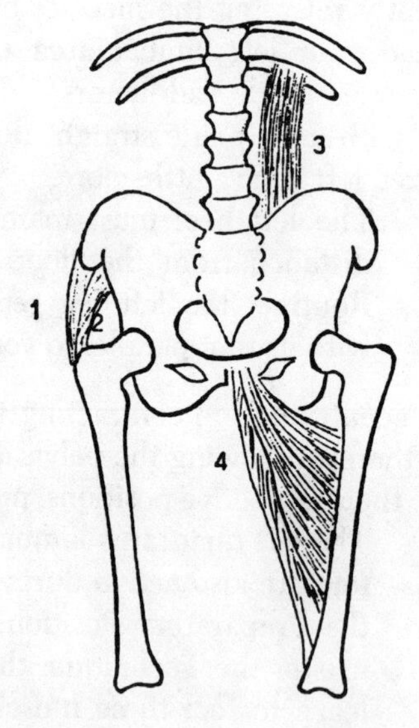

1. gluteus medius } minor glutei
2. gluteus minimus } of the right hip
3. quadratus lumborum
4. adductors (of hip)

Repeat these two phases until your right hip muscles (in particular) have had a good workout. Then resume the Well-balanced Stance by first returning the sole of your left foot to its proper position, below the left hip joint, and subsequently distributing your weight equally over both feet. Then go through the entire Routine standing on your left leg, involving the muscles in your left hip, right lumbar area, and right inner thigh.

This Mensendieck Routine serves to release accumulated strain in the lower back and helps to trim your hips and thighs. Practice the Routine whenever you have to stand and wait; for instance, while waiting for a bus.

NOTE: For an even more complete relaxation of the muscles in your lower back, you can add the forward down tilt of the pelvis (Technique 16C) to the sideward down tilt in phase one of Technique 22; then raise the pelvis in both respects during phase two.

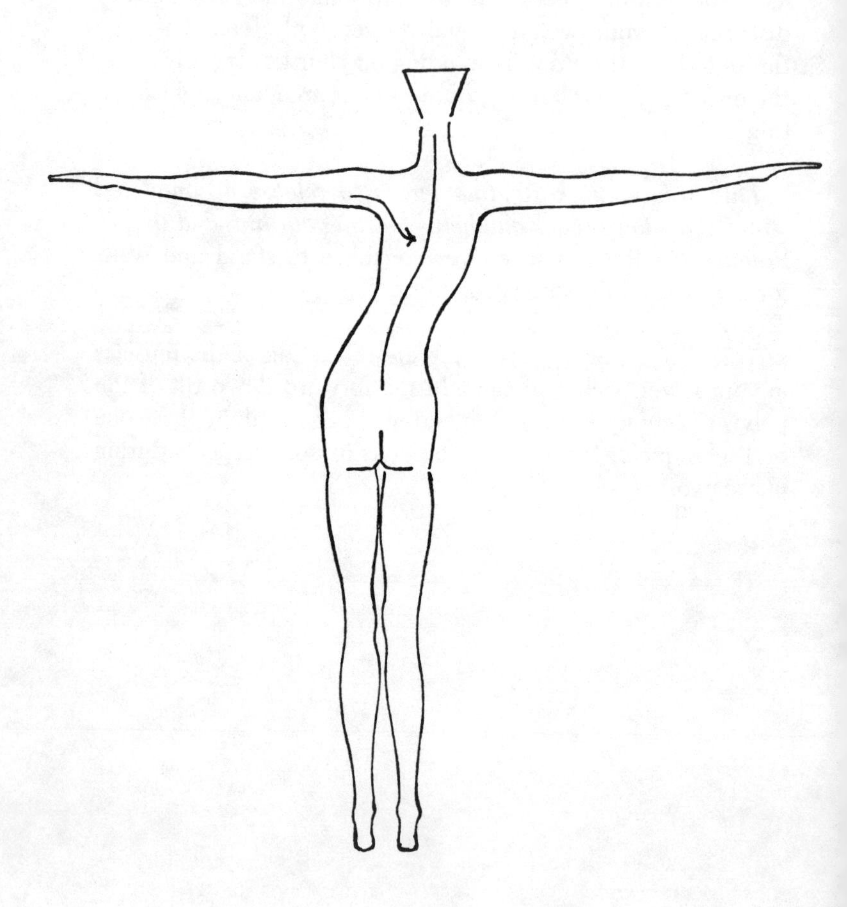

Draw your shoulder blade diagonally downward, then let it push your chest sideways.

Mensendieck Technique 23: CHEST SHIFT

PREPARATORY POSITION: The Well-balanced Stance. Rotate your arms a quarter turn in the shoulder joints, until the palms of your hands face your thighs. Bend both straight hands in the wrist joints, so that your wrists will distance from the thighs. While keeping the shoulder blades low and alongside your spinal column, raise your straight arms exactly sideways in the shoulder joints until your wrists are just above shoulder level. Straighten your hands until they are exactly in line with the arms.

• Exhale, while drawing your left shoulder blade diagonally downward toward your spine; then allow this shoulder blade to push your thoracic vertebrae (thus your chest) as far as possible toward your right.

Maintain your arms frontal, wrists and hands just above horizontal. The shoulder on the pushing side should not be higher than your other shoulder. Keep your pelvis upright and straight above the feet. Watch out that your body does not twist but remains frontal.

• Inhale, while discontinuing the push and returning your chest to its normal position, exactly straight above the pelvis and your feet.

• Exhale, while drawing your right shoulder blade diagonally downward toward your spine. Then allow it to push your chest as far as possible to your left.

Carefully observe the same detailed instruction as stated under phase one.

• Inhale, while discontinuing the push and returning your chest to its normal position, straight above the pelvis and your feet.

Repeat these four phases once more. Finish this Routine as follows: Bend both straight hands upward in the wrists; pull your arms down with the muscles alongside the chest; straighten the hands in line with the arms; rotate your arms inward in the shoulder joints until the palms of your hands face backward.

This Mensendieck Routine aims at releasing accumulated strain in the lower back muscles and at limbering up and straightening out your spinal column.

Next time when you build up the Preparatory Position with your arms try to apply resistance while rotating and raising the arms and while bending your hands in the wrists, as this demands action in muscles that have little to do in your daily life activities. Do the same when moving your arms out of their preparatory position. Counteracting with opposing muscle groups results in moving firmer and slower. Whenever you are instructed to move "firmly," you are meant to counteract with those muscle groups that oppose the movement under discussion.

Is it not strange that notwithstanding honest endeavors of parents and physical educators, the educational mills turn out people in need of posture correction?

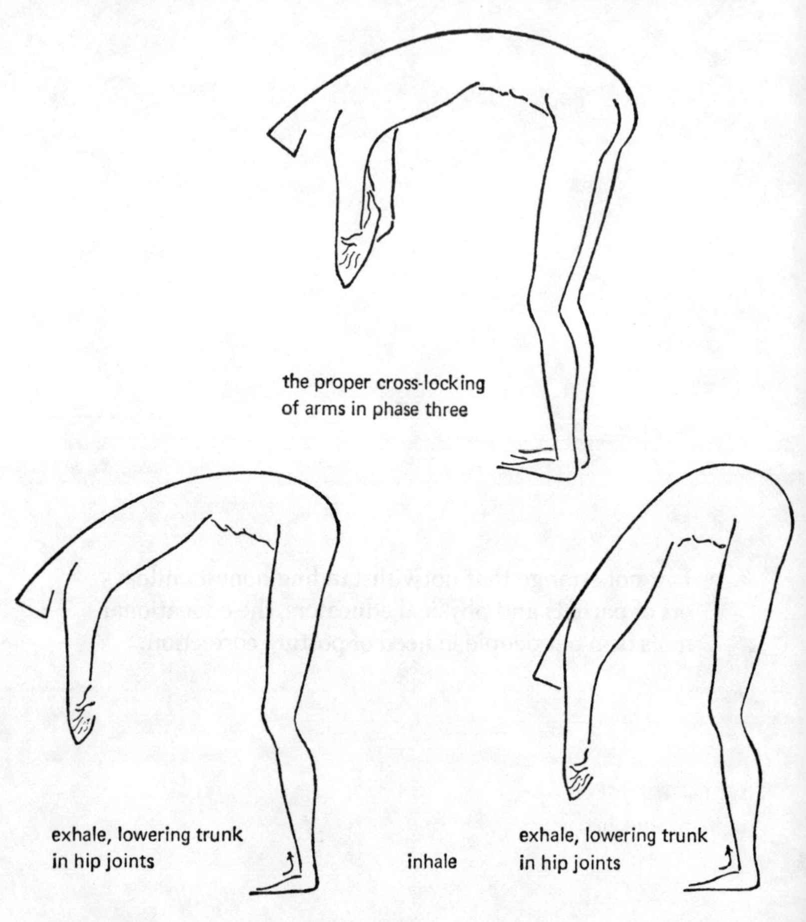

Beginners may not succeed noticeably in achieving such results right away. The muscles tend to be too taut because of their task of holding your body upright most of the day. Without bouncing the trunk, gently invite your muscles to release more of their tension while your trunk is suspended; let your trunk follow the weight of your arms and head down.

Mensendieck Technique 24: FORWARD TRUNK SUSPENSION

PREPARATORY POSITION: The Well-balanced Stance.

• Inhale, while tilting your pelvis forward (phase one of Technique 16C).

Move your tailbone in the direction of the ceiling rather than backward.

• Exhale, while slowly releasing your long back muscles from the sacrum to the back of your head. This will gently lower the upper part of your trunk as well as your neck and head.

Be conscious of the movement of your spine, vertebra for vertebra; eventually the crown of your head will point toward the floor. Maintain your weight toward the balls of your feet, knees slightly bent, throughout the whole Routine.

• Inhale, while crossing your forearms. Then relax your arms in the shoulder and become aware of the sensation of heaviness in arms and head.

If you properly cross the forearms, they will stay locked without your hands holding on to your arms.

• Exhale, while consciously elongating all the muscles along the back of your (slightly bent) knees, your hip joints, and your spine, which will bring about a further lowering of your trunk in the hip joints.

• Inhale and pause.

• Exhale, while again allowing the muscles along the back side of your body to stretch out even more than before . . . to again lower the trunk farther.

• Inhale, while gently dropping your relaxed forearms out of their cross-locked position and keeping your hands relaxed.

• Exhale, while pulling your buttocks under, to raise your pelvis in the hip joints. Then resume the Well-balanced Stance by stacking your vertebrae, from the sacrum upward, gradually drawing your shoulder blades toward the spine and down.

If you have not experienced any dizziness or lightheadedness while executing the Technique, you can lower the suspended trunk more often than twice when you next do it. Each time exhale while smoothly lowering the trunk deeper, and pause while inhaling. Bouncing the trunk is incorrect, because then you would allow the muscles to take back what they gave each time, while the idea is to invite them to give more and more, time and again.

This Mensendieck Routine is a highly effective relaxant to be used between Mensendieck exercises. It is also a great circulation stimulator, facilitating the blood's rushing toward your heart and lungs and brains, organs which the blood has to reach against gravitational pull during your active hours. Whenever you feel drained or otherwise in need of a quick revitalizer, suspend the trunk forward for a while. And, if your shoulder muscles feel tired or stiff, interlock your fingers behind your back; then move your trunk through the suspension Routine while keeping your arms far away from your back, which will bring them overhead when your head points toward the floor. The blood rushing down through your arms usually "soaks" the shoulder muscles enough to clear up the discomfort.

NOTE: Efficient Breathing is the responsibility of specific muscle groups and must be learned. During the breathing-in phase of the Forward Trunk Suspension you can gain distinct awareness of the muscles responsible for expanding your rib cage backward and sideways, as you were instructed to do in phase one of Mensendieck Technique 4.

ARMS RAISE

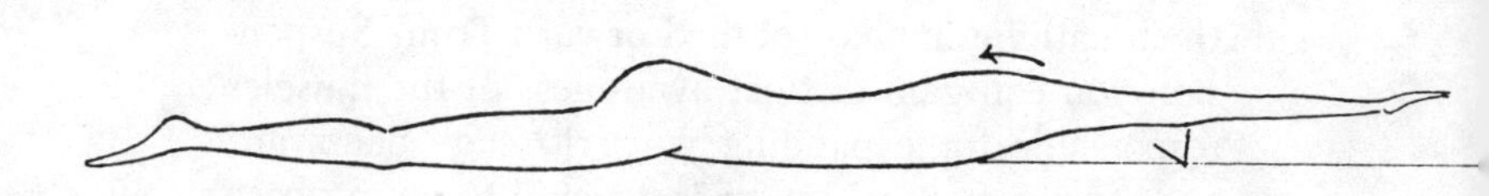

lift your arms (only) off the floor . . . before drawing them sideways

XXII. CONDITIONING YOUR BACK MUSCLES

Mensendieck Technique 25: AIRPLANE
A. Arms Raise
B. Head-and-arms Raise

PREPARATORY POSITION: Lower your body to the floor, as instructed in chapter XII, until you lie prone on your belly, with your head resting on its forehead. Extend your arms alongside your head, on the floor.

A.

• Inhale, while lifting your straight arms off the floor, alongside your head; try to raise them at least four inches, to bring your upper arms level with your ears or higher.

• Exhale, while drawing your high and straight arms sideways until they are at shoulder level.

• Inhale, while returning the high and straight arms toward your ears; try to finish with your arms at least four inches off the floor, close to your head.

• Exhale, while lowering your straight arms to the floor.

Repeat these four phases. Then practice the following:

HEAD AND ARMS RAISE

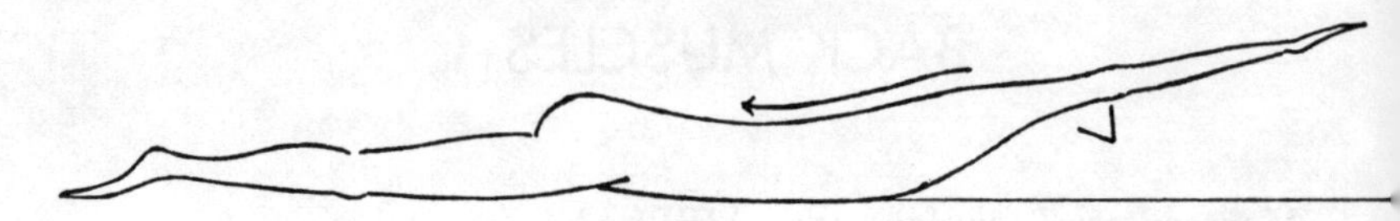

lift your arms together with your neck and head . . . before drawing the arms sideways

B.

• Inhale, while lifting your straight arms as well as your neck, which will take your forehead off the floor.

Try raising the neck first from out its base without the arms, to master this neck motion (without moving the head relative to the neck), then let the arms join in the lift.

• Exhale, while drawing your high and straight arms sideways until they are at shoulder level.

• Inhale, while returning the high and straight arms toward your ears.

• Exhale, while lowering your straight arms to the floor, together with your neck, so that your head will again come to rest on its forehead.

Repeat these four phases a few times. Then get up from the floor the Mensendieck way.

These Mensendieck Routines, as well as the following two, will particularly strengthen your shoulder girdle and back muscles.

If you have a dowager's hump through years of neglecting your lower neck area, often practice the lifting of your neck-head (only) as one unit while lying prone.

If you tried Technique 25 and found that you were hardly able to lift your arms off the floor, practice Techniques 18 and 19 often, until you can succeed in executing the Airplane.

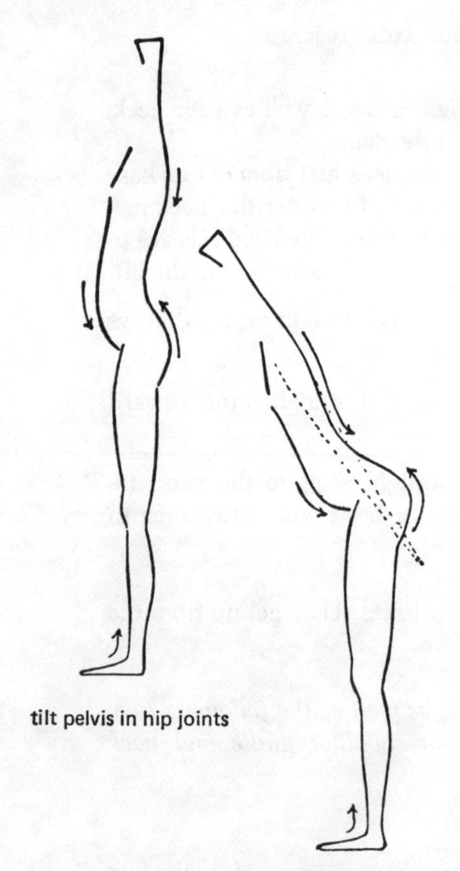

tilt pelvis in hip joints

tilt trunk in hip joints while maintaining your pelvis in forward tilted position and without yielding shoulders and arms to gravity

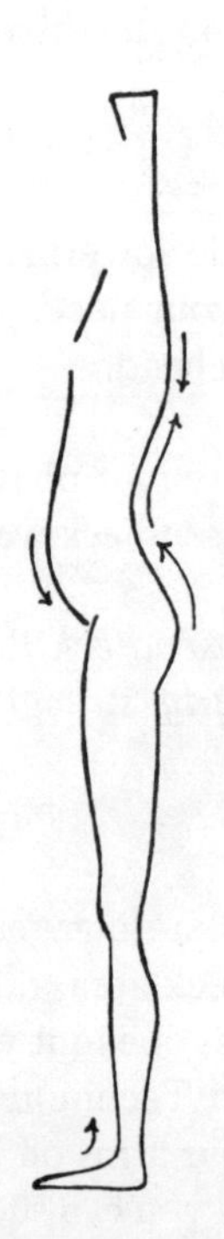

raise upper trunk in lumbar vertebrae (and trunk in hip joints) while maintaining your pelvis in forward tilted position

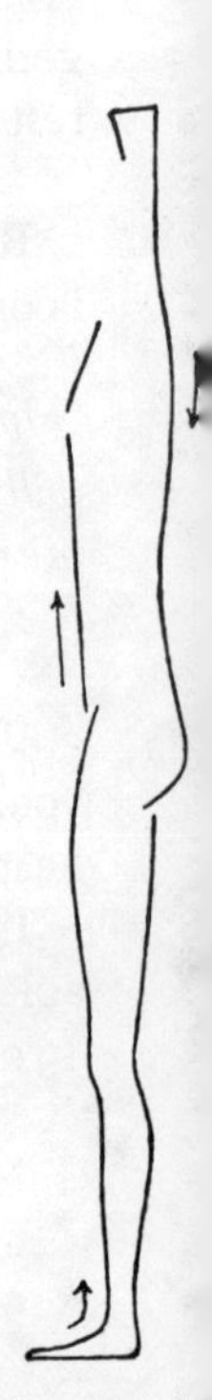

raise pelvis ir hip joints

Mensendieck Technique 26: STRAIGHT FORWARD TRUNK BEND

PREPARATORY POSITION: The Well-balanced Stance. Secure your shoulder blades well down and toward the spine. Keep your arms straight with the upper arms close to the shoulder blade, to prevent yielding to gravity in the coming exercise.

• Inhale, while slowly releasing the tension in your buttocks and abdominal muscles, thereby allowing your pelvis to tilt forward in the hip joints so that your lumbar lordosis becomes more pronounced (as instructed in Technique 16C).

Keep your knees slightly bent and your weight toward the balls of your feet.

• Exhale, while releasing your buttocks even more, thereby allowing your whole trunk to slowly bend straight forward in the hip joints (only) until the trunk, with the lumbar section of your back still concave, is at an angle of about 45 degrees. The whole trunk will tilt forward properly if you think of moving your tailbone upward. Keep your weight well forward, knees slightly bent, upper arms alongside the shoulder blades, shoulder blades toward the spine and down, your neck-and-head in line with the spine and tall to crown.

• Breathe once in and out, with your trunk stationary in oblique position.

• Inhale, while elevating the upper trunk relative to your pelvis, with the muscles in your lower back, thereby increasing the lumbar lordosis. Actually, your whole trunk will raise in its hip joints, but it's better not to concentrate on that. If you keep in mind raising your sternum, it will help to properly elevate the (upper) trunk together with your neck-and-head (in line with the spine).

• Exhale, while drawing your buttocks together and under and your abdominal muscles upward from groin to sternum, thereby allowing your pelvis to be raised in the hip joints until your lumbar lordosis is again shallowly concave. At the conclusion of this phase, you must have resumed the Well-balanced Stance.

Repeat these five phases several times, staying well forward in your feet, knees slightly bent. Once this Technique is mastered, try the Pelvis Rocking while your trunk is oblique:

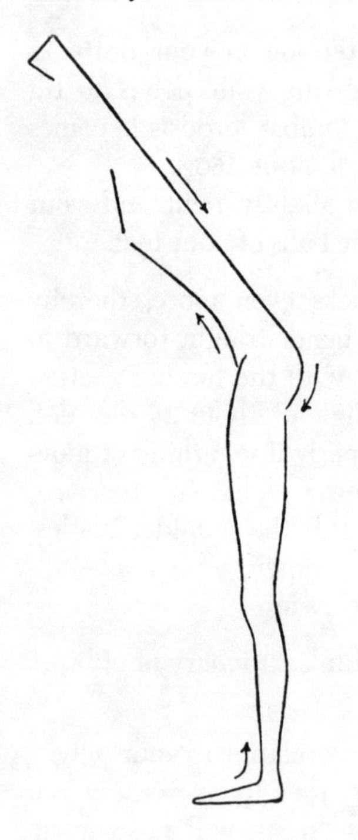

raise your pelvis out of its forward tilt until your spine is as straight as a ruler

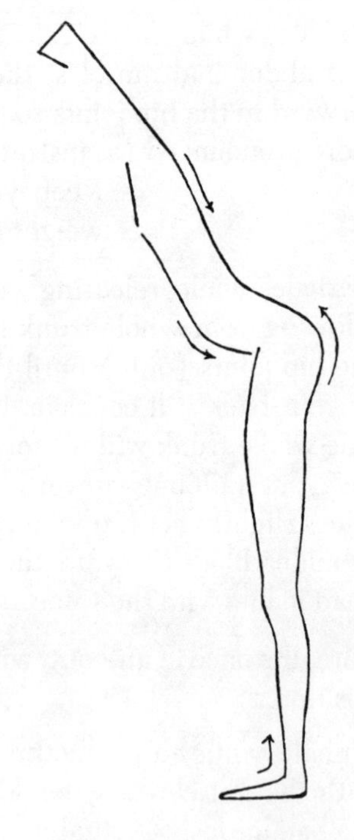

tilt your pelvis way forward again to resume the preparatory position

Mensendieck Technique 27: PELVIS ROCKING IN THE STRAIGHT TRUNK BEND

PREPARATORY POSITION: The Well-balanced Stance. Then execute phases one and two of Technique 26 to assume the oblique trunk position with your lower back concave. Throughout the Routine you must keep your weight well forward, your knees slightly bent, your arms straight, upper arms secured alongside the shoulder blades, the shoulder blades toward the spine and down, your neck-and-head in line with the spine and tall to crown.

• Exhale, while drawing your buttocks together and under and your abdominal muscles upward from groin to sternum, thereby allowing your pelvis to be raised in the hip joints until your spinal column is as straight as a ruler.

Maintain your shoulder blades and arms in their proper positions, while eliminating the lumbar lordosis.

• Inhale, while releasing the tension in your buttocks and abdominal muscles, thereby tilting the pelvis way forward in the hip joints, which will re-establish the lumbar lordosis. Check your shoulder blades and arms at the conclusion of this phase; pull them decisively to where they should have stayed all the time.

Repeat these two phases several times, before resuming the Well-balanced Stance through the execution of the last two phases of Technique 26. Following this, practice the Forward Trunk Suspension (Technique 24) in order to release possible leftover strain in your back muscles.

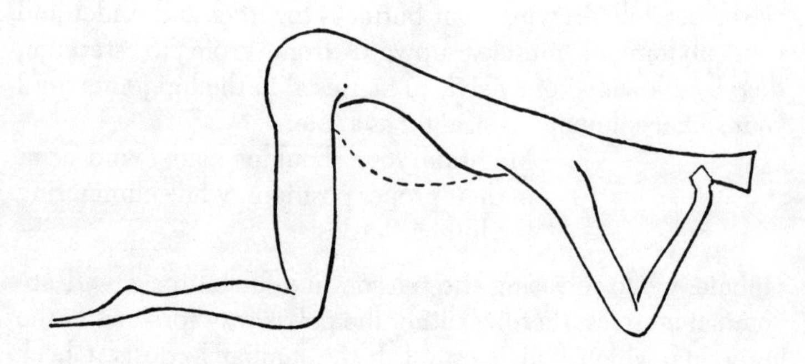

After "filling your belly with air," shape it like a greyhound's from groin to chest.

XXIII. PROPER ABDOMINAL CONTRACTION

Mensendieck Technique 28: BELLY BALLOON

Preparatory Position: On your knees, with the feet extended straight back on the floor. While keeping the upper legs vertical, place your elbows under the shoulders on the floor. Support your head with your hands, wrists under chin. Neck tall to crown.

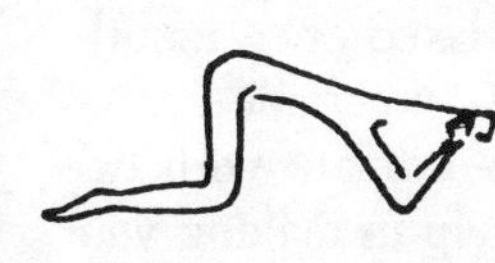

• Inhale, while allowing the abdominal wall to balloon out toward the floor.

Think of filling your belly with air.

• Exhale, while contracting the muscles in your groin first, then follow the contraction through toward your belly button, and finally all the way toward your sternum. Think of shaping your belly like a greyhound's.

No bony part of your body may move. In this Technique, the buttock muscles may not assist your belly muscles but must remain completely relaxed.

Repeat these two phases several times. Once the Technique is mastered, execute the same while sitting cross-legged in tailor-fashion with your back straight so that your sternum is secured (forward up). Then execute the Technique while sitting well balanced on a stool and eventually while in the Well-balanced Stance.

This Mensendieck Routine is ideal for acquiring skill in proper contractions from groin to chest. During our active hours, gravity continually pulls our abdominal wall down; hence continual counteraction is required. You can't practice this Routine often enough, whenever you sit or stand and have a moment to attend to your belly muscles. It will help to unlearn the common tendency to contract the muscles at the stomach level only. *The Routine is also an excellent reconditioner following childbirth or abdominal surgery.*

Remember always to exhale through a relaxed open mouth while practicing Mensendieck. If you haven't experienced earlier how proper abdominal contractions co-ordinate with exhaling, the Belly Balloon will be of great help in making you conscious of this fact. Since we are here reviewing some of the instructions presented earlier in this book, did you kneel down and did you get up from the floor as instructed in chapter XI? Try to always apply what you have learned before. Always move into and out of the required Preparatory Position the Mensendieck way. This will be conducive to habitually correct movement patterns throughout your daily activities.

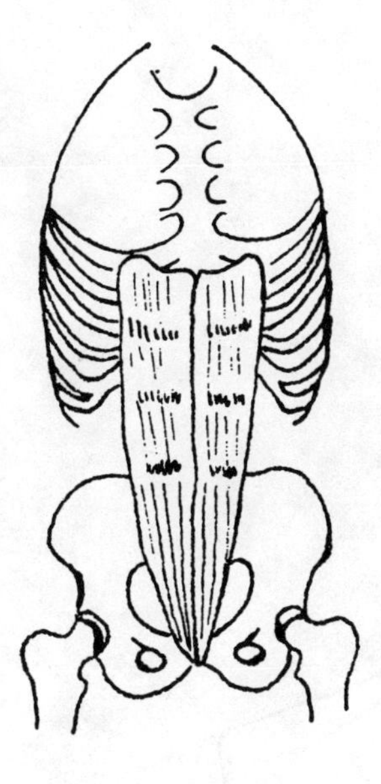

Rectus abdominis (belly) muscles, most responsible for hooking up pubic bones to sternum when your body is upright.

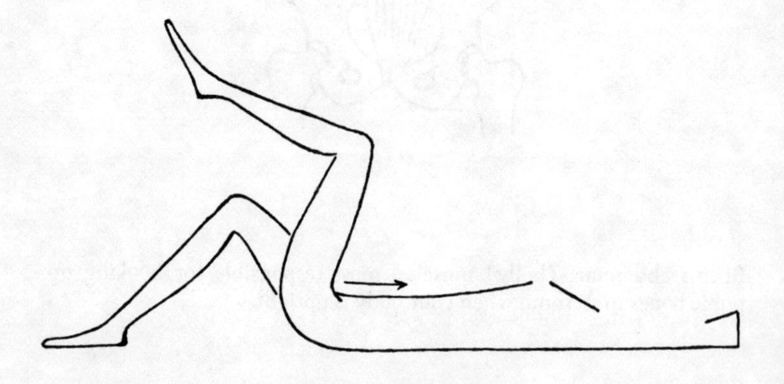

Raise or lower leg in hip joint without releasing your abdominal contraction in the slightest, to stabilize your lumbar vertebrae against the floor which will protect your back.

XXIV. CONDITIONING YOUR ABDOMINAL MUSCLES

Mensendieck Technique 29: PROGRESSIVE LEG RAISING
A.B.C. Raising Bent Legs
D.E.F.G. Raising Straight Legs

PREPARATORY POSITION: Techniques 29A through 29C. Supine, with the arms alongside your body, palms down. Draw your knees up, one at a time, while exhaling and contracting the abdominal muscles, until the soles of your feet and your lumbar vertebrae are on the floor. Shoulders low; shoulder blades well toward the spine; your neck tall to crown.

A.

• Inhale gently through your well-opened nostrils, as you should always do.

• Exhale, while (1) slowly drawing the pubic bones toward the sternum with your abdominal muscles to stabilize the pelvis upright, i.e., with lumbars on the floor, and then (2) slowly raise one bent leg, in its hip joint, straight upward toward your chest, without releasing your abdominal contraction in the slightest.

• Inhale.

• Exhale, while (1) concentrating again on the steady upward pull of your abdominal muscles, so that your pubic bones cannot possibly distance from your sternum and your lumbars cannot possibly leave the floor, and then (2) slowly lower the raised bent leg in its hip joint until your foot is back on the floor next to your other foot, without releasing your abdominal contraction in the slightest.

Repeat these four phases with the other bent leg. Continue to practice the Technique with one leg at a time, to indelibly etch in your mind the exact sequential order in which legs must be raised and lowered, in order to protect your back from harm.

This Mensendieck Routine aims at making you aware of how abdominal muscle activity is essential for the stabilization of your lumbar vertebrae against the floor when legs are being raised. No abdominal muscles are attached to your legs; therefore there is not one abdominal muscle directly involved in the raising of legs. Their involvement can and should be indirect, by "hanging on" to your pubic bones when the weight of your raised legs tend to tilt the pelvis forward. This tendency also results from another factor: the major psoas muscle runs from the bodies of your lumbar vertebrae through your pelvis to the top of your leg. This muscle is directly involved in raising a leg. It will also pull your lumbars toward your leg if you fail to keep them well anchored against the floor with your belly muscles.

NOTE: As soon as lumbar vertebrae are allowed to leave the floor during raising of legs, your belly protrudes and its muscles are being stretched. This is also why it cannot be stressed enough that you should never lift more leg weight than your abdominal muscles can cope with. Bent legs are easier to lift than straight legs, thus continue to practice raising bent legs until you can afford the luxury of raising straight ones. With this information in your mind, proceed through the following increasingly demanding Techniques in this chapter, to gradually condition your abdominal muscles through months of practicing.

The best procedure during each exercise session is to begin your leg raises with Routine 29A in which the least weight has to be lifted off the floor. Then go on to the next leg raise Routine and the next and so on, in the order given, until you have reached your limit. Just make sure that you have challenged your belly muscles to that very limit before passing on to other Routines.

B.

- Inhale.
- Exhale, while (1) drawing pubic bones toward sternum and (2) slowly raising one bent leg and then the other, without releasing your abdominal hold.
- Inhale.
- Exhale, while drawing pubic bones toward sternum and slowly lowering the bent leg which was raised first, and then the other, without releasing your abdominal hold.

Repeat these four phases in reverse order, essential for the balancing of your musculature. In other words, if you first started with your left leg and finished with your right, you must now start with your right leg and finish with your left.

C.

• Inhale.

• Exhale, while drawing pubic bones toward sternum and then slowly raising both bent legs simultaneously . . . if your abdominal muscles can prevent your lumbar vertebrae from leaving the floor. Otherwise start with one leg and follow with the other as soon as you can be sure that your abdominal muscles can cope with the load of both bent legs.

• Inhale.

• Exhale, while drawing pubic bones toward sternum, and then slowly lowering both bent legs simultaneously—up to the point that you feel your lumbar vertebrae are about to be pulled off the floor by the weight of your legs—which may be all the way. If the latter is not the case, continue the downward movement with one leg until it rests on the floor, and only then follow through with the other leg.

Repeat these four phases until your abdominal muscles have had a good workout.

If you intend to straighten the legs at the conclusion of this Routine, utilize this opportunity to give the springs in your foot arches a workout: Caterpillar (Technique 13) simultaneously with both feet until it is no longer possible. Then slide both heels over the floor while exhaling and emphasizing a lower abdominal upward pull.

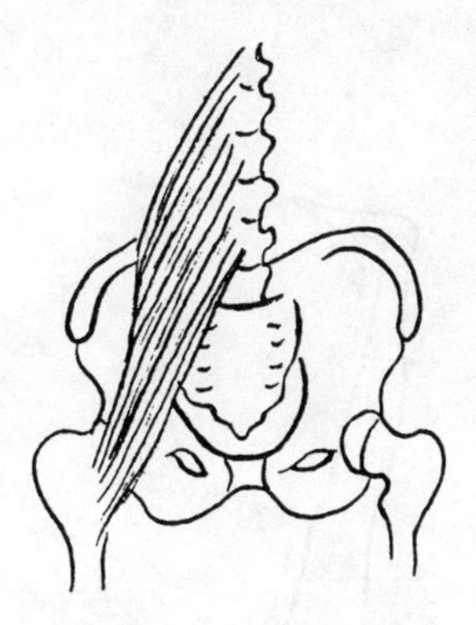

The right psoas major as seen from the front. This muscle must not be allowed to pull your lumbars off the floor when legs are being raised, so do not allow the distance between pubic bones and sternum to increase, which would result if you released your abdominal hold on your pubic bones.

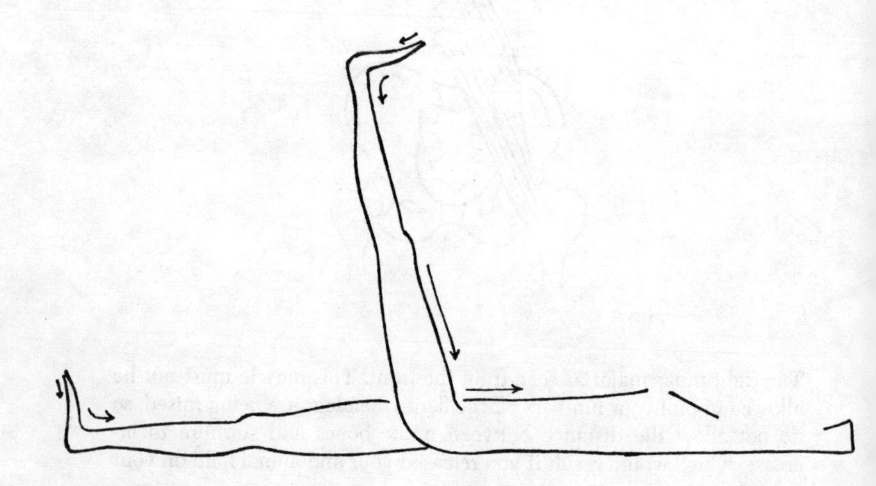

Carrying the feet well balanced helps to keep your knees straight.

Preparatory Position: Techniques 29D through 29G. Supine, with your legs straight in line with your trunk, your feet in well-balanced position (Technique 9). In this straight-legged supine position your lumbar vertebrae will not quite touch the floor.

Rest your arms alongside your body, palms down, shoulders low, shoulder blades well toward your spine; your neck tall to crown.

D.

• Inhale.

• Exhale, while drawing pubic bones toward sternum and then slowly raising one straight leg in its hip joint.

• Inhale.

• Exhale, while drawing pubic bone toward sternum and then slowly lowering the straight raised leg, in its hip joint, to the floor.

Repeat these four phases with the other straight leg. Never move your straight legs any higher than you possibly can with feet carried well balanced and knees well extended. Hyperextension of knee joints will not occur when the legs are raised. In fact, gravity tends to make your knees bend, hence you must decisively contract the extensors of your knees (quadriceps femoris) along the front of your upper legs during the raising and lowering of legs.

E.

• Inhale.

• Exhale, while drawing pubic bones toward sternum and then slowly raising one straight leg and then the other. This must bring about an immediate touching of the floor with your lumbars.

• Inhale.

• Exhale, while drawing pubic bones toward sternum and then slowly lowering the straight leg which was raised first, and then the other. Try to keep your lumbars on the floor as long as possible.

Repeat these four phases with your legs in reversed order. From this point on, you may have to check your neck regularly during leg raises, maintain it tall to crown, thus your neck muscles relaxed.

F.

• Inhale.

• Exhale, while drawing pubic bones toward sternum and then slowly raising one straight leg and then the other (as in 29E). This must bring about an immediate touching of the floor with your lumbars.

• Inhale.

• Exhale, while drawing pubic bones toward sternum and then slowly lowering both straight legs simultaneously — for as long as your abdominal muscles can keep your lumbar vertebrae down. Subsequently continue the downward movement with one leg and then the other.

Repeat these four phases for as long as your abdominal hold can protect your lower back; be sure to keep your neck tall. Raise your legs in reverse order each time.

G.

• Inhale.

• Exhale, while drawing pubic bones toward sternum and then slowly raising both straight legs simultaneously, or start with one and let the other one follow as soon as your abdominal muscles can properly counteract the load of both legs, feet in well-balanced position.

• Inhale.

• Exhale, while drawing pubic bones toward sternum and then slowly lowering both straight legs simultaneously, finishing with one at a time toward the end, if necessary, to protect your back.

Repeat these four phases until your whole body has had a good workout.

These basic Mensendieck leg-raise Routines are powerful abdominal conditioners.

Quadriceps femoris muscle of right leg, front view. The muscle runs from your pelvis over the front of your hip joint, thigh bone, and knee joint to the top of your lower leg; thus it can raise your leg forward in the hip joint, while keeping your knee straight!

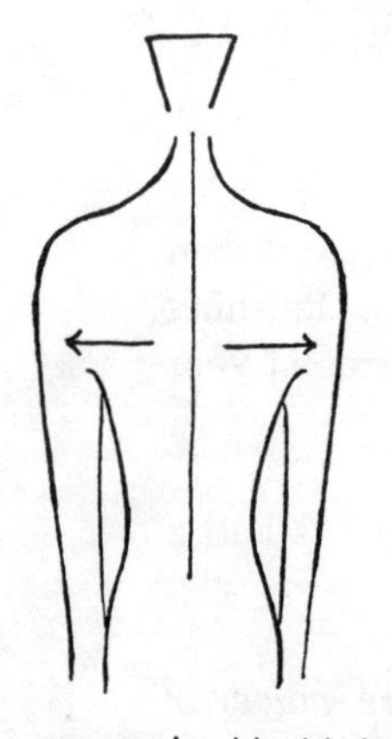

separate shoulder blades

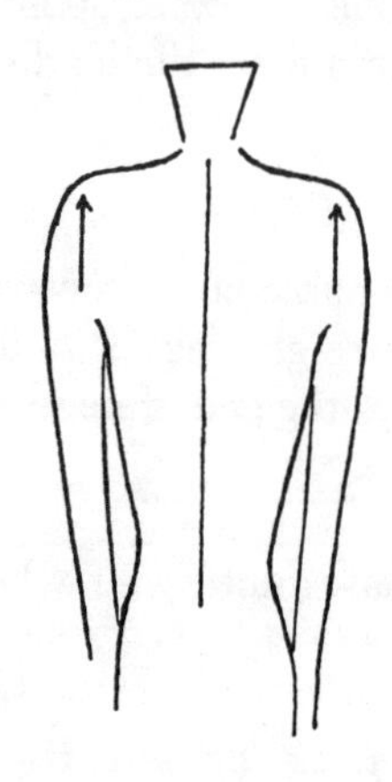

raise them

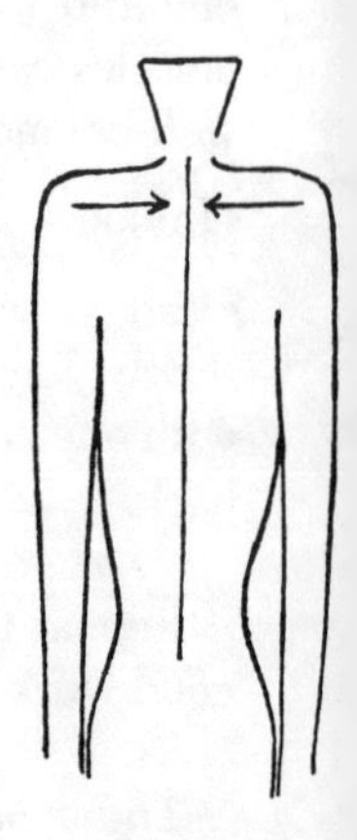

pull them to spine

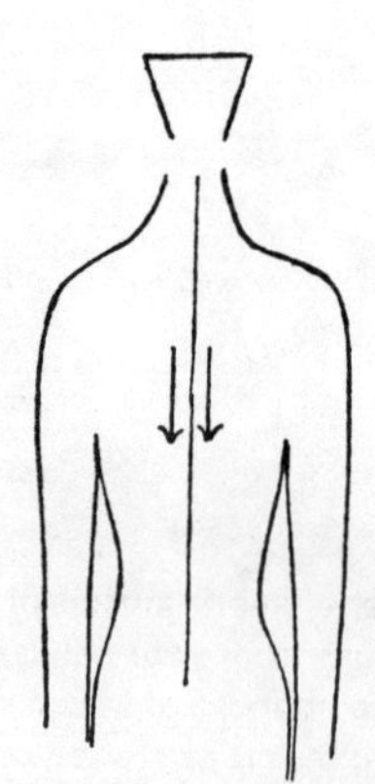

while exhaling, draw shoulder blades closely along spine downward . . . all the way;(remember to contract firmly your belly muscles while you exhale, otherwise your upper trunk will be pulled backward in this phase)

XXV. CONDITIONING THE MUSCLES OF YOUR SHOULDER GIRDLE

Mensendieck Technique 30: TRAPEZIUS EXERCISE

PREPARATORY POSITION: The Well-balanced Stance. Your arms straight, alongside your body, palms facing backward. Concentrate on the inner margins of your shoulder blades, alongside your spine.

• Inhale, while pulling these inner margins sideways, away from your spine, thereby allowing your shoulders to come forward; then draw these inner margins way upward toward your ears; subsequently draw these inner margins toward your spine.

• Exhale, while slowly and firmly drawing these inner margins downward, closely along your spine, until your shoulder blades have resumed their proper position.

Be careful not to compress your body during this phase; keep your pelvis upright, your sternum forward and up, your neck tall to crown.

Repeat these two phases several times, while keeping the arms straight alongside your body. Your thumbs must stay in touch with the "trouser-seam." When you practice this Technique for the first time, skip the exact breathing co-ordination, until you are familiar with the fourfold movement of your shoulder blades. Each shoulder blade describes a square, the first three sides of which are necessary preparatory motions for the principal phase (two) of the Technique, i.e., the fourth side of the square.

This Mensendieck Routine particularly aims at strengthening the lower part of the trapezius muscles, whose task it is to counteract the tendency of the shoulder girdle to suspend from its articulation with the sternum, as explained in chapter I. In addition, the sliding of your shoulder blades through layers of active muscles is a highly effective shoulder-strain relaxant.

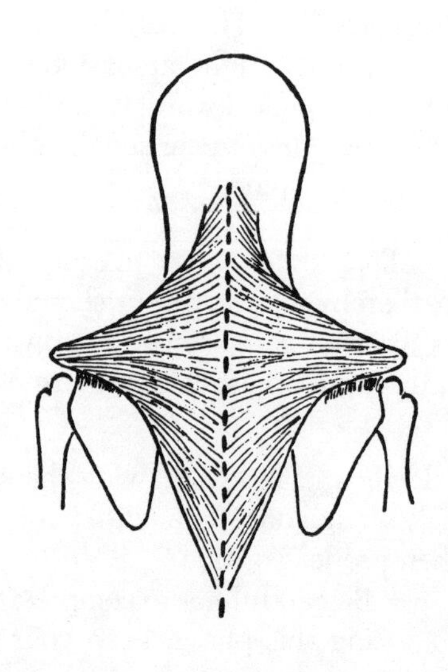

Trapezius muscles

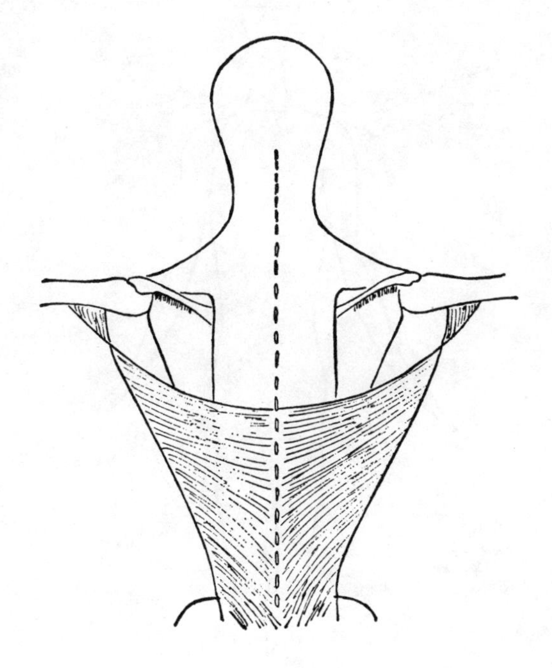

Latissimus muscles

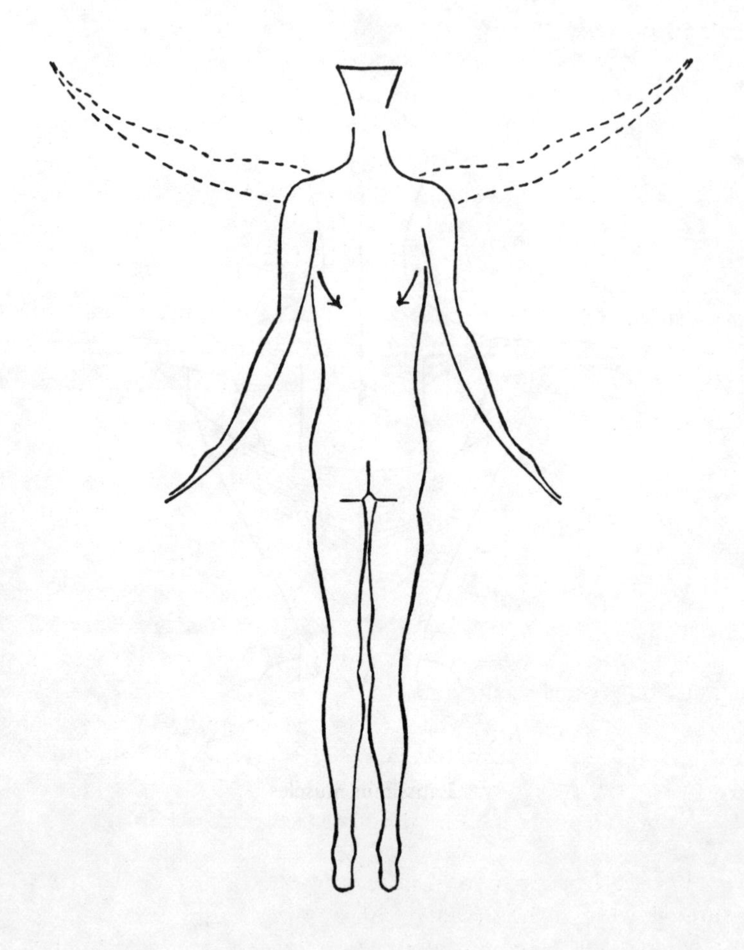

Inhale and raise your rolled-out arms until your straight fingers point diagonally upward, then pull your arms firmly down until your straight fingers point diagonally downward while exhaling.

Mensendieck Technique 31: LATISSIMUS EXERCISE

PREPARATORY POSITION: The Well-balanced Stance. Rotate the upper arms outward in the shoulder joint, then follow this motion through with your forearms until the palms of your straight hands face sideways. Bend your elbows and wrists ever so lightly. The arms have to be maintained in this slightly curved shape throughout the exercise.

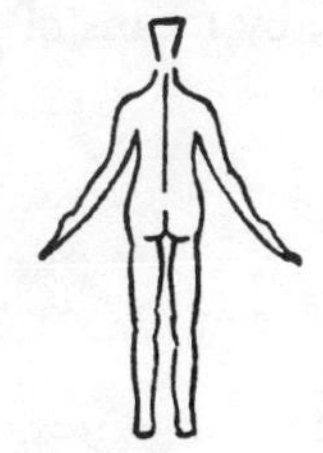

• Inhale, while slowly raising your outward-rotated arms sideways, in the shoulder joints (only), until the upper arms are just above shoulder level, i.e., until your straight fingers point diagonally upward.

Maintain your shoulder girdle low, the inner margin of the shoulder blades next to the spine and parallel with it, snug against the ribs, throughout the exercise.

• Exhale, while firmly contracting the latissimus muscles at the side of your chest, thereby pulling your arms exactly sideways slowly down in the shoulder joints until your straight fingers point diagonally downward.

If you succeed in keeping your slightly curved arms properly outward-rotated, you will not be able to see the surface of your palms in the front mirror image.

Repeat these two phases several times. Then rotate the upper arms inward in the shoulder joints until the fold of the elbows is facing your body. Subsequently follow this rotation through with your forearms until the palms of your straight hands face backward. You must now have resumed the Well-balanced Stance.

This Mensendieck Routine particularly aims at strengthening the latissimus muscles whose main task is to sustain posture. Its contraction "squeezes" your chest, thus your sternum, forward and up. A well-developed latissimus will also help to keep the lower tip of your shoulder blades snug against the ribs and will assist in carrying the shoulder girdle low, by means of its attachment high to the arm.

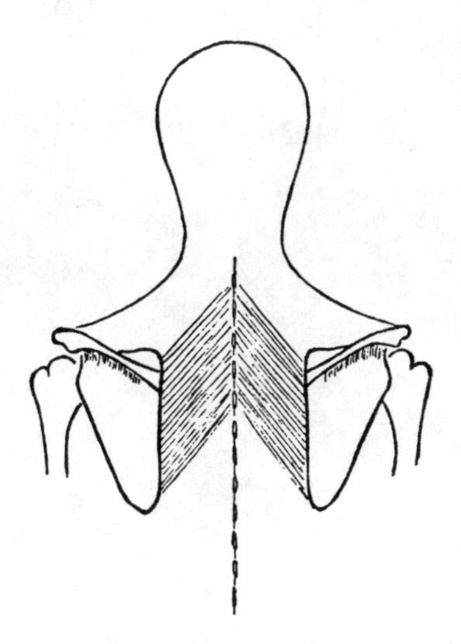

Rhomboideus muscles

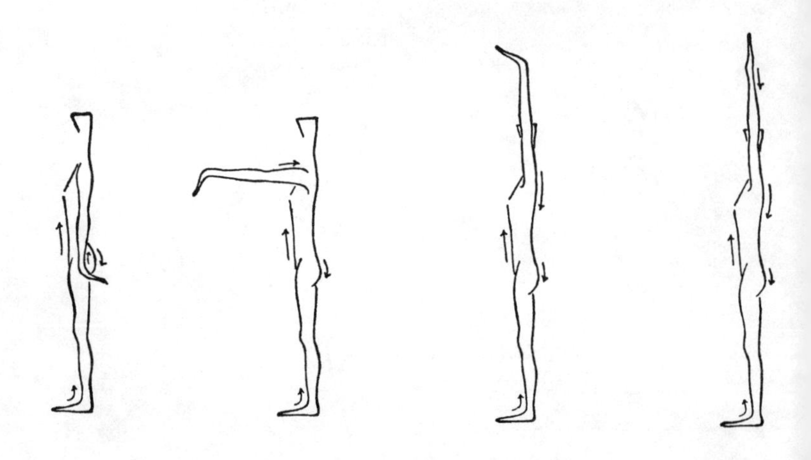

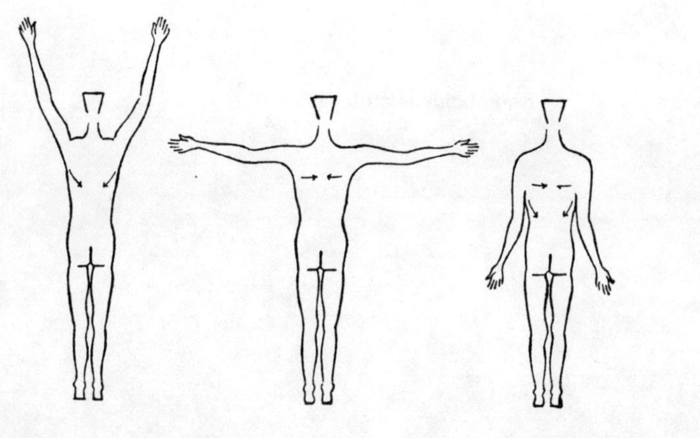

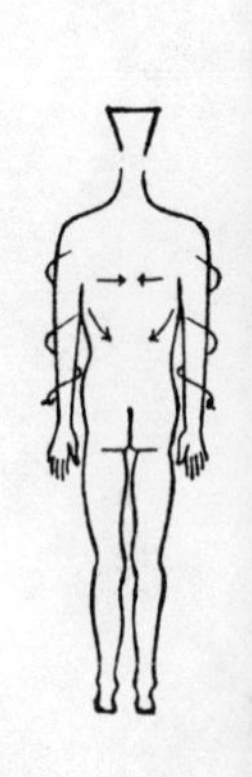

Mensendieck Technique 32: RHOMBOIDEUS EXERCISE
A. In Nine Phases
B. In Two Phases

PREPARATORY POSITION: The Well-balanced Stance. Your arms straight, alongside your body, palms facing backward. Concentrate on the inner margin of your shoulder blades, alongside your spine.

A.

• Bend both straight hands backward in the wrists to acquire a distinct sensation of the end of your arms and to activate usually neglected forearm and hand muscles.

The all-important end-of-extremity sensation will assist your directional awareness.

• Inhale, while firmly raising both arms in the shoulder joint straight forward to shoulder level, thereby keeping your shoulder blades stationary and low, their inner margin next to the spine and parallel with it.

The upper trunk must not compensate backward during the forward elevation of your arms. Maintain correct body alignment throughout the Routine.

• Exhale, correcting your body alignment if necessary.

• Inhale, while firmly raising your arms from shoulder level to alongside your ears, or to as high as your correct body alignment allows.

The lower tip of your shoulder blades have automatically distanced from your spine during this phase, a movement which you were meant to allow reluctantly. The whole shoulder girdle has been raised and tilted somewhat; your arms have rolled out.

• Straighten your hands exactly in line with the arms. Your palms now face forward.

• Exhale, while contracting the latissimus muscles at the sides of your chest, thereby pulling your shoulder girdle down, which will start a sideway-and-down movement of your arms. Then contract the rhomboideus muscles from spine to inner margin of shoulder blades, and pull the lower tips of your shoulder blades low and toward the spinal column, which will pull your arms farther sideways and down until they are at shoulder level. Your palms are still facing forward.

Direct your arms exactly sideways during this phase, i.e., in the frontal plane through the shoulder joints.

• Inhale, correcting your body alignment if necessary.

• Exhale, while firmly contracting the latissimus muscles at the side of your chest, thereby drawing the arms from shoulder level to alongside your body. This movement takes place in the shoulder joints.

• Rotate the upper arms inward in the shoulder joints until the folds of the elbows are facing your body. Then follow this rotation through with your forearms until the palms of your hands face backward. You will now have resumed the Well-balanced Stance.

The inner margin of your shoulder blades should remain next to the spine and parallel with it, low and snug against the ribs, during the inward rotation of your arms.

Repeat these nine phases, while maintaining correct body alignment, until you have fully mastered the Technique. From then on you can co-ordinate the Technique as follows:

B.

• Inhale, while firmly bending both straight hands backward in the wrists, then firmly raising both arms straight forward and on upward until (ideally) they are next to the ears. Then firmly straighten both hands in line with the arms.

Every step of this phase must be executed exactly as you have learned earlier. The same holds true for the following phase.

• Exhale, while firmly pulling the shoulder girdle down, then firmly pulling the lower tip of the shoulder blades toward your spine, then firmly pulling the arms from shoulder level sideways to alongside your body, then firmly rotating your arms inward from shoulder to fingertips.

Repeat these two phases often. The Technique is a marvelous ventilator.

This Mensendieck Routine particularly aims at strengthening the rhomboideus muscles, whose main task is to keep the inner margin of your shoulder blades snug against the ribs alongside the spinal column.

Through practicing the Rhomboideus Exercise you actually experience what we touched upon in chapter XV:

1) How the shoulder blades, thus the shoulder girdle, can remain stationary in correct posture-sustaining position during all arm movements below shoulder level, i.e., during most of your daily life activities;
2) How the whole shoulder girdle is forced out of its correct posture-sustaining position when the arms have to reach above shoulder level, due to the fact that no further motion of the arms relative to the shoulder girdle is possible (in the shoulder joints);

3) Why it is essential to secure the whole shoulder girdle back in its correct posture-sustaining position, as soon as the arms return from above shoulder level, in order to prevent the shoulder girdle from suspending itself from its articulation with the sternum, which would result in a stoop-shouldered posture, as was explained in chapter I.

The main function of all three muscles discussed in this chapter is to sustain posture in perfect co-ordination with one another. The lower part of the trapezius holds the shoulder blade low and toward the spine. The rhomboideus holds the inner margin of the shoulder blade toward the spine as well as snug against the ribs. The latissimus holds the lower tip of the shoulder blade snug against the ribs as well as the whole shoulder girdle low. In other words, the common task of these three muscles is the correct positioning of your shoulder girdle. Succeeding in this is essential for the correct alignment of your thoracic vertebrae, and hence for the forward-and-up carriage of your sternum. This, in turn, is essential for effective functioning of your abdominal muscles and for the effective spreading of your ribs while breathing, as already pointed out in chapter I.

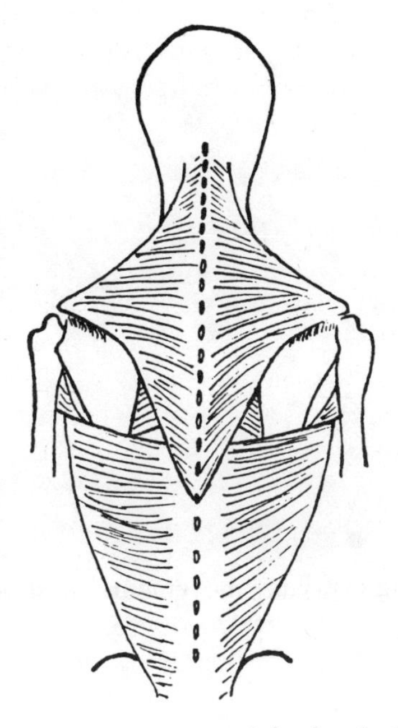

The trapezius covers most of the rhomboideus.

Performing everyday movements constructively preserves energy.

XXVI. HEAD CARRIAGE AND MOVEMENTS

As soon as you succeed in correctly superposing all your body masses from the feet up to the head, the latter will automatically end up perfectly poised, which means that your shoulder and neck muscles hardly have to work in order to hold the head up. Your head is mostly a bony mass and weighs about one fifteenth of your total body weight. A faulty postural alignment, upon which the load of your head cannot rest balanced of its own accord when the trunk is upright, is the main reason for tense and tired muscles commonly found in the neck and shoulder area.

In addition to generally correct posture, in which the chest together with the neck-head can best be carried as a unit, the following special movements of your neck and your head relative to one another can be practiced to prevent, or to release, a neck and shoulder strain. In addition, they will reverse the trend to a double chin.

Mensendieck Technique 33: THE NOD-TURN-TILT

A. Nod
B. Turn
C. Tilt

PREPARATORY POSITION: Sit cross-legged, as for Technique 17A (or on a stool, hands on thighs). Execute the four phases of that Technique to secure a perfectly straight and vertical spine. Then concentrate on your neck from its base upward, all the way to the crown of your head.

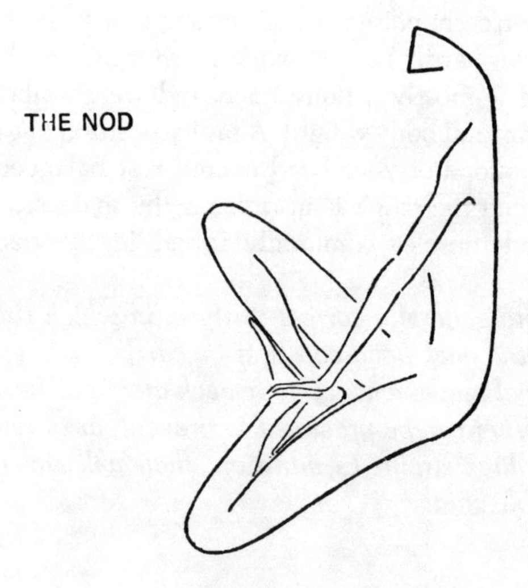

Curve your neck only

A.

• Slowly bend your neck straight forward in its seven vertebrae, starting with the one at the bottom, then the second, and the third, and the fourth, and the fifth, and the sixth with the seventh, then stop. During this whole phase your chin approaches your forward-and-up sternum.

Be careful to curve your neck only, keeping your back absolutely straight and vertical.

• Slowly raise your head in the neck vertebrae, again starting their return to normal alignment with the lowest neck vertebra, then the one higher up, and so on until your tall-to-crown spine has resumed its ideal shape.

The head should not be raised relative to the neck. Instead raise your neck relative to the thoracic vertebrae, which will result in the lifting of your head.

Repeat these two phases several times, slowly and consciously; your neck and shoulder muscles must be stretched, not tensed. Then execute the four phases of Technique 17A to first relax your long back muscles from the sacrum to the back of your head, and then to realign your spinal column from tail to crown. With your spine straight and vertical, practice the following Technique.

THE TURN

Revolve head on axis through neck vertebrae to tall crown of head.

B.

• Imagine an axis through your neck vertebrae to the tall crown of your head.

On this axis your head is going to revolve. Thus try to remain very much aware of this axis throughout the following four phases.

• Slowly turn your head toward your left and feel the neck muscles being stretch-twisted around the tall-to-crown axis, while your straight back remains stationary.

Your spine below the neck must not rotate.

• Slowly turn the head back until you are facing straight forward again, with the dimple in your chin straight above your jugular notch in the front mirror image.

The jugular notch is the dent immediately above your sternum, between the articulations of the collar bones with the sternum.

• Slowly turn the head toward your right and feel the neck muscles being stretch-twisted around the tall-to-crown axis, while your straight back remains stationary.

Your spine below the neck must not rotate.

• Slowly turn the head back until you are facing straight forward again, chin dimple above jugular notch.

Repeat these five phases several times, slowly and consciously; your neck and shoulder muscles must be stretched, not tensed. Then execute the four phases of Technique 17A, to first relax your long back muscles from the sacrum to the back of your head, and to then realign your spinal column from tail to crown. With your spine straight and vertical, practice the following:

THE TILT

Think of axis through neck vertebrae to tall crown of head and incline the top sideways.

C.

- Slowly incline your tall-to-crown head (only) sideways toward your left, while maintaining your chin dimple exactly above your jugular notch and your spinal column straight and vertical.

- Slowly raise your tall-to-crown head out of its oblique position, while maintaining your chin dimple exactly above your jugular notch and your spinal column straight and vertical.

- Slowly incline your tall-to-crown head (only) sideways toward your right, while maintaining your chin dimple exactly above your jugular notch and your spinal column straight and vertical.

- Slowly raise your tall-to-crown head out of its oblique position, while maintaining your chin dimple exactly above your jugular notch and your spinal column straight and vertical.

Repeat these four phases several times, slowly and consciously; your neck and shoulder muscles must be stretched, not tensed. Then practice Technique 17A a few times to release possible strain in your back muscles from holding your spine erect for quite some time. Get up from the floor the Mensendieck way, to give other muscles in your body an advantageous workout.

Mensendieck Routine 33 teaches you how to carry your head straight, with the chin at right angle to your throat. Through this Routine you are made distinctly aware of the neck-tallness-to-crown that must continually be pursued in correct posture. As soon as you allow your spine to lose its tallness to the crown of your heavy head, the load of the latter will tend to compress not only your neck but the whole vertebral column.

After having practiced each section of Technique 33 a number of times, try to slowly practice all eleven phases consecutively. Pause briefly each time upon the return of your head to its perfectly upright face forward position and be conscious of the tallness to crown sensation. Subsequently, practice the whole Routine with your eyes closed; open them briefly each time, after the head has carefully resumed its upright face forward position according to your perception, in order to visually check whether or not you perceived the correct sensation with your eyes closed.

XXVII. SWINGS

Unnecessary residual tensions in your muscles can be released through various swings, of which three Routines are given here. Make it a point to habitually practice a swing at the end of a strenuous day or at the conclusion of a Mensendieck training session.

Initiate each swing by allowing a particular part of your body to move forward and down while yielding it to gravity and exhaling. It will thus gain momentum at first, then lose its momentum, and finally come to a spontaneous halt at the end of its range of possible movement.

Immediately following this instant of minimal pause, allow the part of your body to move in reverse while yielding it to gravity and inhaling. It will thus gain momentum at first, then lose its momentum, and must finally come to a spontaneous halt in the exact position it was in at the start of the swing.

Always aim for regrasping a firm hold on that particular part of your body, which moved freely through the forward-down swing and its reversal, at the point where it has resumed its original position and follow your inhaling through in full. Then, again release your hold and repeat the swing, forward-down while exhaling and forward-up while inhaling, completing the inhalation during the hold at the conclusion of the entire swing.

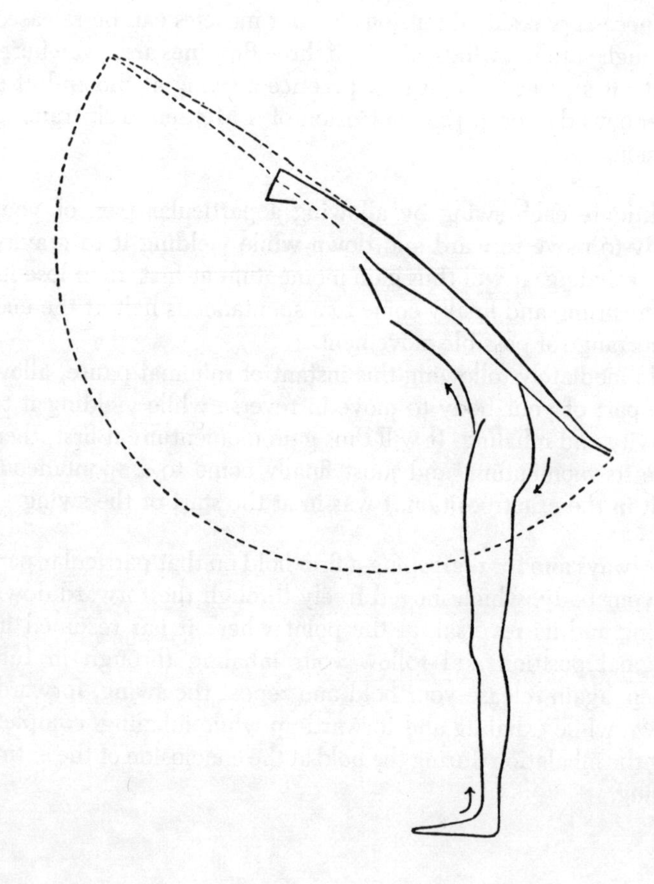

Swing your arms freely till your hands are behind your hips . . . and back till your arms have returned to their original position alongside your ears.

Mensendieck Technique 34: THE ROUND FORWARD TRUNK BEND SWING

PREPARATORY POSITION: Round Forward Trunk Bend, with the arms next to ears. Start out from the Well-balanced Stance. Next:

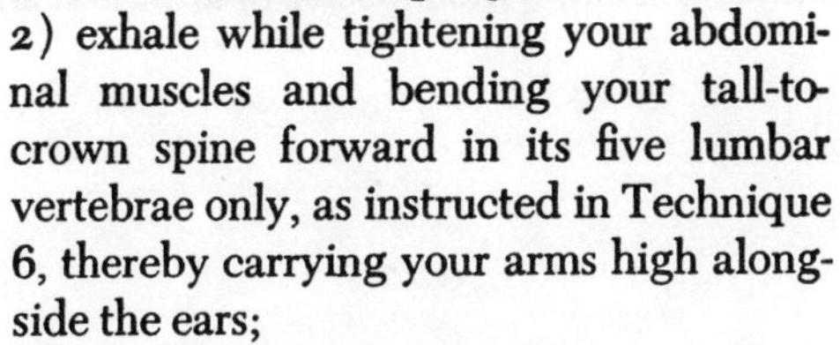

1) inhale while raising your arms forward and up until they are next to the ears as instructed in Technique 32B;

2) exhale while tightening your abdominal muscles and bending your tall-to-crown spine forward in its five lumbar vertebrae only, as instructed in Technique 6, thereby carrying your arms high alongside the ears;

3) inhale and think of keeping your body weight well forward toward the balls of the feet, with your knees slightly bent, throughout the Routine.

• While exhaling, release the hold on your arms and allow them to move freely forward-down in the shoulder joints until your hands are behind your hips.

Your trunk should remain in perfect round bend position.

• While inhaling, allow your arms to move freely forward-up until they have resumed their original position alongside your ears.

Your trunk is still in perfect round bend position.

• Continue to inhale until your lungs are fully expanded, while your body remains stationary in the resumed Preparatory Position.

Repeat these three phases two more times, while keeping your body weight well forward toward the balls of the feet. Subsequently, exhale, while raising the upper trunk together with neck-head and arms-next-to-ears, until your tall-to-crown spine has resumed its ideal erect shape and place, as instructed in Technique 6. Finally, slowly lower your arms sideways, as instructed in Technique 32B, and rotate them inward until your palms face backward.

This Mensendieck Routine particularly aims at releasing residual tension in the muscles of arms and shoulder girdle.

Graceful movements are recognized by everybody.

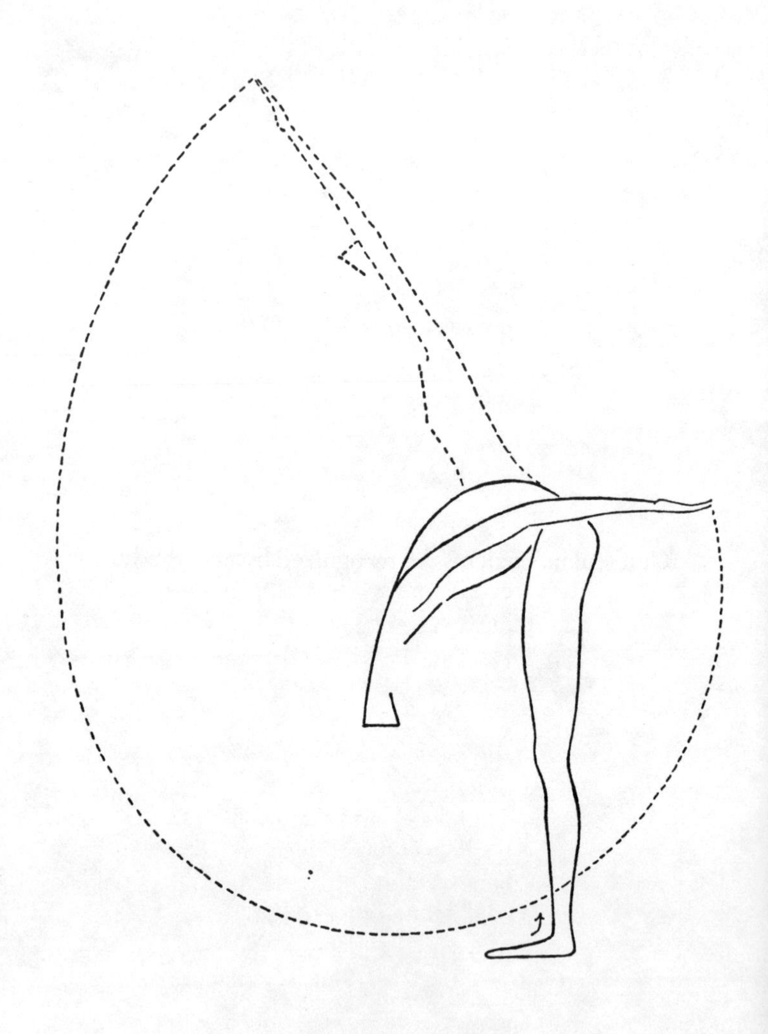

Be sure to keep your neck tall to crown in line with the rest of your spine while swinging down and returning to your Preparatory Position.

Mensendieck Technique 35: THE STRAIGHT FORWARD TRUNK BEND SWING

PREPARATORY POSITION: Straight Forward Trunk Bend, with the arms next to ears. Start out from the Well-balanced Stance. Next:

1) inhale, while raising your arms forward and up until they are next to the ears as instructed in Technique 32B;
2) exhale;
3) inhale, while slowly releasing the tension in your buttocks and abdominal muscles, thereby allowing your pelvis to tilt forward in the hip joints as instructed in Technique 16C;
4) exhale, while releasing your buttocks even more, thereby allowing the trunk together with neck-head and arms-next-to-ears to slowly bend straight forward in hip joints until the trunk is at an angle of about 45 degrees, with your lower back still slightly hollow, as instructed in Technique 26;
5) inhale and think of keeping your body weight well forward toward the balls of the feet, with your knees slightly bent, throughout the Routine.

• Exhale, while releasing the muscles in your lower back, thereby allowing your upper trunk together with neck-head and arms to gently drop deep down in its lumbar vertebrae. The momentum gained should properly swing your arms way backward until your hands are behind your hips.

• Inhale, while gently swinging your arms and upper trunk together with neck-head forward and up in your shoulder joints and lumbar vertebrae respectively, until you have resumed the Preparatory Position.

Make sure that your arms are actually returned next to the ears and that your lower back is again slightly hollow.

• Continue to inhale until your lungs are fully expanded.

Repeat these three phases two more times, while keeping your body weight well forward toward the balls of the feet. Subsequently, inhale while raising the upper trunk together with neck-head and arms-next-to-ears, relative to your still forward-tilted pelvis, thereby increasing your lumbar lordosis; then exhale, while raising the pelvis in its hip joints until your tall-to-crown spine has resumed its ideal erect shape and place, as instructed in Technique 26. Finally, slowly lower your arms sideways, as instructed in Technique 32B, and rotate them inward until your palms face backward.

This Mensendieck Routine particularly aims at releasing residual tension in the muscles of the lower back as well as in those of arms and shoulder girdle.

Correct posture is the ultimate result of intelligent multiple muscle control.

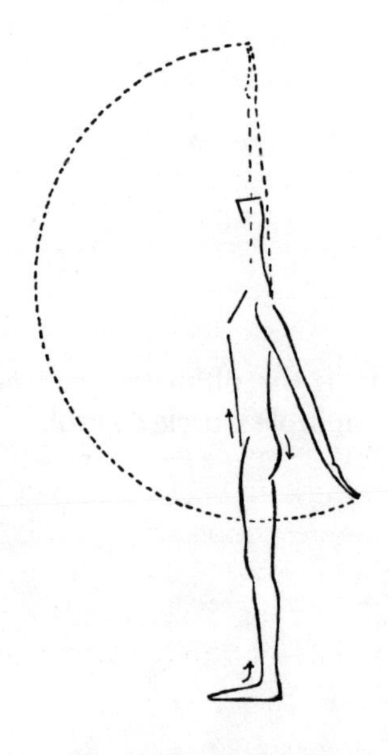

A. swing your arms only

Mensendieck Technique 36: PROGRESSIVE SWINGS

A. Arms Only
B. Head and Arms
C. Upper Trunk and Head and Arms
D. Whole Trunk and Head and Arms
E. Legs and Trunk and Head and Arms

PREPARATORY POSITION: Standing with the arms next to ears. Get into that Position as follows:

After having carefully constructed the Well-balanced Stance, inhale while raising your arms forward and up until next to the ears as instructed in Technique 32B. This Position has to be correctly resumed at the conclusion of each of the following swings. During each swing, you should not move any part of your skeleton except those participating in the particular swing. Keep your body weight toward the balls of the feet throughout the Routine.

A.

• While exhaling, freely swing your arms forward-down in the (low) shoulder joints until your hands are behind your hips.

• While inhaling, freely swing your arms forward-up until they have resumed their original position alongside your ears.

• Continue to inhale until your lungs are fully expanded.

Repeat these three phases two more times, then proceed:

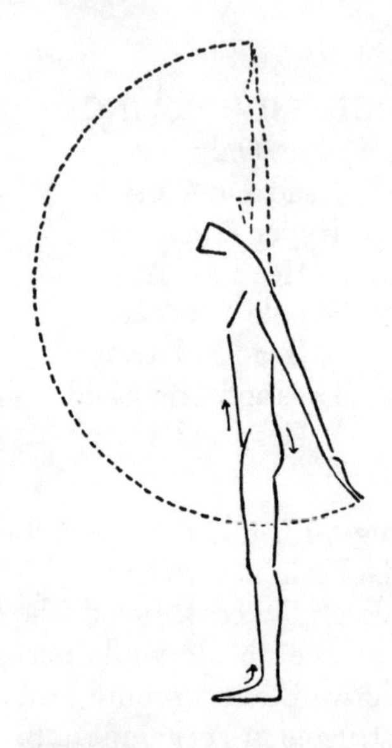

B. neck and head take part in the movement

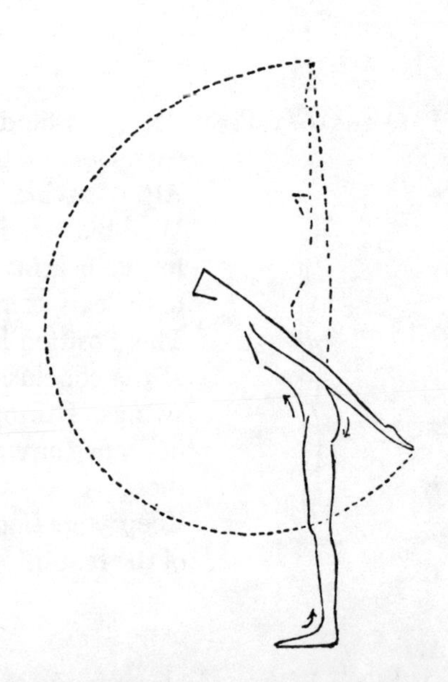

C. upper portion of body takes part in the movement

B.

• While exhaling, bend your head forward in the neck vertebrae and freely swing your arms forward-down until your hands are behind your hips.

• While inhaling, raise your head in the neck vertebrae and freely swing your arms forward-up until you have resumed the Preparatory Position.

• Continue to inhale until your lungs are fully expanded.

Repeat these three phases two more times, then proceed:

C.

• While exhaling, bend your upper trunk together with neck-head forward in the lumbar vertebrae and freely swing your arms forward-down until your hands are behind your hips.

• While inhaling, raise your upper trunk together with neck-head and freely swing your arms forward-up until you have resumed the Preparatory Position.

• Continue to inhale until your lungs are fully expanded.

Repeat these three phases two more times, then proceed:

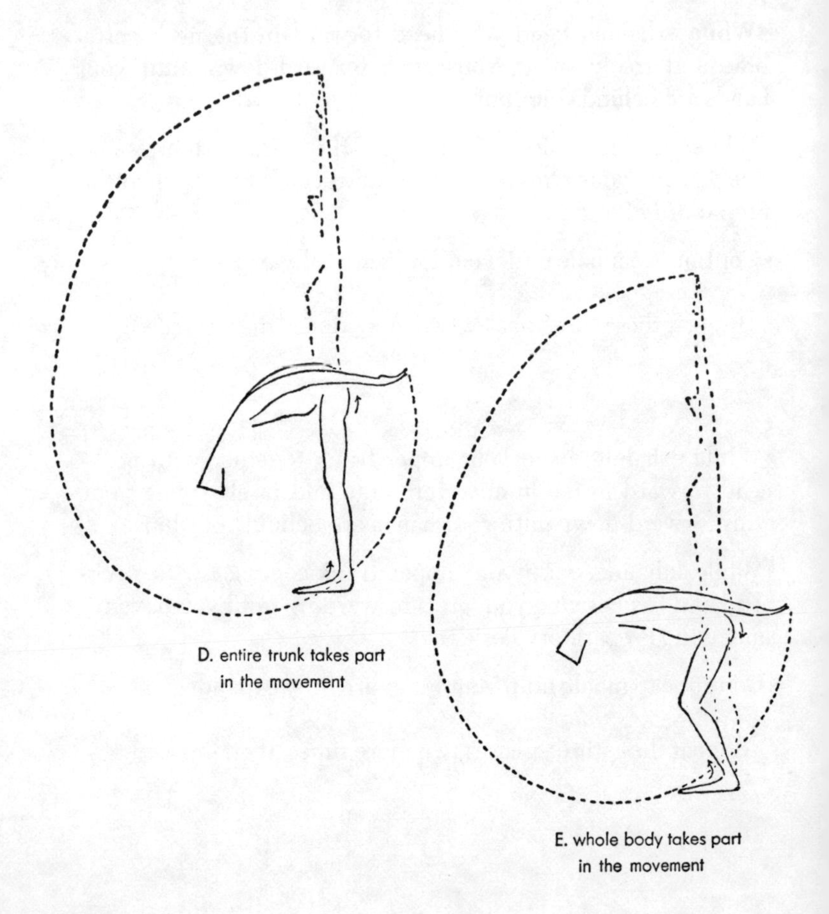

D. entire trunk takes part in the movement

E. whole body takes part in the movement

D.

• While exhaling, bend your trunk together with neck-head forward in the hip joints and freely swing your arms forward-down until your hands are behind your hips.

• While inhaling, raise your trunk together with neck-head and freely swing your arms forward-up until you have resumed the Preparatory Position.

• Continue to inhale until your lungs are fully expanded.

Repeat these three phases two more times, then proceed:

E.

• While exhaling, bend your knees as well as your trunk together with neck-head forward in the hip joints and freely swing your arms forward-down until your hands are behind your hips. Straighten the knees slightly during the last part of this swing, and bend them again at the very beginning of the ensuing forward-up swing. This makes the movement of your knees rhythmic and bouncing.

• While inhaling, bend, then straighten your knees when you raise your trunk together with neck-head and freely swing your arms forward-up until you have resumed the Preparatory Position.

• Continue to inhale until your lungs are fully expanded.

Repeat these three phases two more times, then finish the series of swings by allowing your arms (only) to freely swing forward-down, while remaining completely relaxed in and around your shoulder joints. This will automatically result in several consecutive arm swings, back and forth, and a continual decrease in the range of the pendulumlike movement. Eventually your arms will come to rest alongside your body in perfect Well-balanced Stance.

This Mensendieck Routine stimulates your respiration and circulation and releases residual tension throughout your body.

NOTE: Those among you who feel like swinging on, at the conclusion of 36E, can first repeat once each of the five stages of the Technique in reverse order, before finishing with the consecutive arm swings. This will challenge your selective muscle control. When you can correctly and smoothly execute the five stages of the Progressive Swings consecutively, you have succeeded in mastering adequate selective muscle control, which is the main purpose of Mensendieck training and of this book.

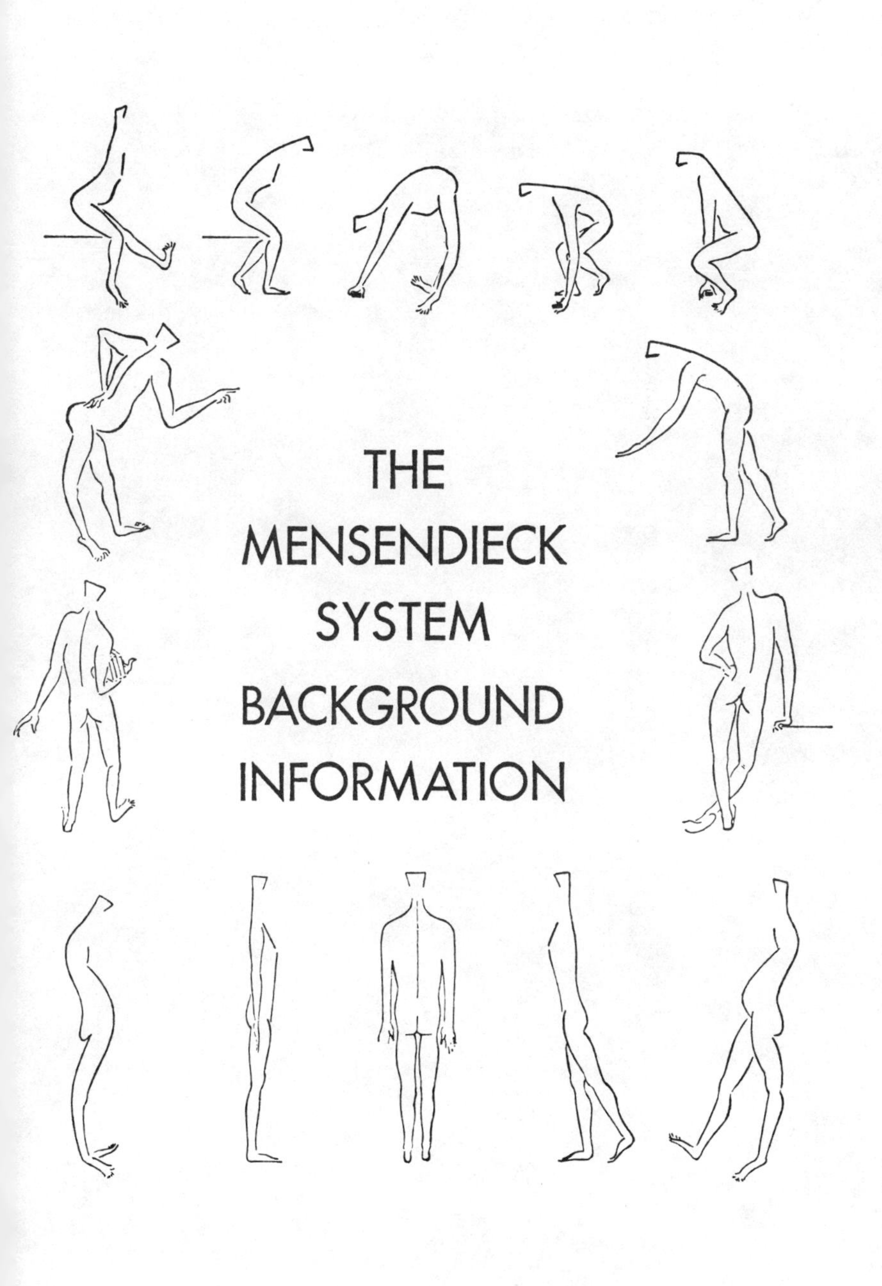

THE MENSENDIECK SYSTEM BACKGROUND INFORMATION

In the late 1800s, while studying at the Academy of Sculpture in Paris, Bess M. Mensendieck, American by birth, found that she would rather sculpture human bodies than inanimate matter; being confronted with live models whose bodies were supposedly the "cream of the crop," she became increasingly concerned about their poor neuromuscular co-ordination which resulted in ungraceful postures and flabby muscles; also it struck her as odd that so many of them were suffering aches and pains.

She left Paris to study medicine in Zurich, Switzerland, and became one of the world's first female physicians.

Equipped with a great deal of basic knowledge regarding the human body, Dr. Mensendieck returned to Paris around the turn of the century and obtained a position at its Medical School where, for years, she probed and experimented in the field of kinesiology until she knew exactly what every human being would have to know and do to master proper neuromuscular co-ordination in order to cope with the mechanics of his body. On the basis of her research, Dr. Mensendieck designed her system of functional body education and returned to the United States in 1905, to introduce "The Mensendieck System." Unfortunately for this country, Dr. Mensendieck was prevented from doing so due to the puritanical Comstock laws which were then in effect. Because of these laws, American publishers would have had to risk spending one year in jail and paying a fine of one thousand dollars if they should publish her books containing illustrations of unclothed bodies, which illustrations were indispensable for the clarification of her work.* Hence, Dr. Mensendieck returned to Europe where, in Germany, she was first given the opportunity of presenting her contribution to her fellow men. In the 1920s, the publishing company F. Bruckmann A.-G., München, published her first **four books, to wit: *Korpercultur des Frau; Bewegungsprobleme; Functionelles Frauenturnen*, and *Anmut der Bewegung im Taglichen Leben* translated into IT'S UP TO YOU. Two Mensendieck Associations were founded: The Deutsche Mensendieck Bund (in Hamburg), and the Mensendieck-Bund (in München). However, it is said that German therapists became involved and because they interfered with the carefully designed Mensendieck System, Dr. Mensendieck withdrew her approval and support.**

In 1929, she affirmed her withdrawal in writing, in the Dutch translation of the book *Functionelles Frauenturnen* (*Functioneele Lichaams-Oefeningen voor de Vrouw en het Kind*, H. J. Paris, Amsterdam). As a result, Mensendieck work, after a very promising start, failed in Germany.

On July 29, 1930, the International Mensendieck League was founded at Larvik, Norway; the Seat of the League was established in The Hague, the Netherlands. It consisted of the Norsk Mensendieck-Forbund, founded in Norway in the year 1920; the Dansk Mensendieck Forbund, founded in Denmark in the year 1925; the Svenska Mensendieck-Forbundet, founded in Sweden in the year 1926; and the Nederlandse Mensendieck Bond, founded in the Netherlands in the year 1928.

The objective of the International Mensendieck League was the protection of The Mensendieck System against dilution and/or mutilation. In order to preserve the significant effectiveness of the Mensendieck System in its pure and unadulterated form, Dr. Mensendieck established the following system of "Checks and Balances": Candidate Mensendieck specialists must follow Mensendieck training at a nationally accredited

* *The Mensendieck System of Functional Exercises*, Foreword, Bess M. Mensendieck, Southworth-Anthoensen Press, 1937. This book is available in most university libraries.

Mensendieck Institute; certification of Mensendieck specialists — after passing all prescribed examinations and other graduation requirements — shall only be furnished by one of the national Mensendieck Associations, specifically authorized by Dr. Mensendieck to do so.†

After having thus set up her Mensendieck stronghold in the northwestern part of the European continent, in exact accordance with her remarkable vision regarding its significant potentials for unfoldment, Dr. Mensendieck returned to New York and, in private practice, devoted the rest of her lifetime to making her pupils aware of the methodical way in which they should maneuver their "body-mobile" correctly throughout all everyday activities for the purpose of maintaining it in tiptop condition. Meanwhile, the Comstock laws had been repealed and she could finally have books published in her own country: *It's Up to You*, J. J. Little & Ives Co., 1931; *The Mensendieck System of Functional Exercises*, Southworth-Anthoensen Press, 1937; *Look Better, Feel Better*, Harper & Brothers, 1954.

A few years before she passed away in 1957, Dr. Mensendieck returned to Europe, at the age of ninety-four, to conduct a seminar in Denmark. When she was complimented regarding her zest for life and her exceptional vitality, she replied: "It is because of the spreading of my ribs."

Although the Mensendieck profession is a highly respected one throughout the world, there are as yet very few Mensendieck specialists practicing outside Europe. This is because the demand for their services is so great in Europe. Also, the enrollment at any of the European Mensendieck Institutes is limited to approximately two dozen freshmen per year, due to

† At the time of this writing, Mensendieck is recognized and regulated by law as a completely separate paramedical profession in Norway and the Netherlands; efforts to obtain similar legal status are progressing in Denmark and Sweden. Since the 1920s, a few thousand Mensendieck specialists have graduated from the European Mensendieck Institutes.

the intensive individual instruction each student receives daily throughout the first two years of schooling. This daily intensive individual instruction is not only required for the purpose of attaining mastership of the numerous basic Mensendieck Techniques (thereby perfecting the student's own posture and movement habits), but, in the same way as the experiencing of personal psychoanalysis is essential for the future psychoanalyst, Mensendieck specialists-to-be must experience the re-education of the muscles themselves, so they may acquire the skill of diagnosing the postural and movement problems of future pupils (those with no particular medical problem) and patients (medical cases, referred by a physician) in a competent manner.

The daily personal experiencing of Mensendieck over an extended period is essential to the acquisition of professional competency. It is for this reason that practitioners of other allied disciplines, such as graduate physical therapists, must follow the complete Mensendieck curriculum in order to achieve that competency. In addition, the theory of all Mensendieck Techniques and the theory of the unique Mensendieck teaching method are an integral part of the Mensendieck curriculum from the very start. For the physical therapist, in particular, the transformation into a competent Mensendieck specialist may indeed prove to be an excruciatingly difficult process. Generally, as is the case with practitioners in any (a)vocation, excellence depends a very great deal on the basic make-up of the particular individual attempting to attain perfection in his particular field of endeavor. Characteristically different basic potentials are required in an individual for the eventual unfolding of a successful physical therapist, as opposed to those potentials required in an individual to become a successful Mensendieck specialist.

The awareness of the diametrical difference in the professional services rendered (physical therapy primarily from the "outside-in" as opposed to the "inside-out" approach of Mensendieck) stimulated the respective European governments, in which the accredited Mensendieck Institutes are established, to cause the academic training for these two professions to be totally independent of one another. In many European medical schools, candidate physicians are indoctrinated in the distinction between the service a Mensendieck specialist can provide on the one hand, and a physical therapist on the other, so that they can later intelligently evaluate which of the two should be prescribed for the maximum benefit of a particular patient.

In the Netherlands, during the third academic year, the candidate Mensendieck specialist must successfully serve at least three different full-time internships at various teaching hospitals and rehabilitation centers. Before being permitted to begin her first internship, the student must pass written and oral examinations in numerous study subjects, all of which are compulsory. Finally, and before receiving her diploma at the end of the internship year, the student must take an oath similar to the Hippocratic oath required for physicians.

The Mensendieck curriculum is complex. For the interested reader, the examination requirements for the Mensendieck study in the Netherlands are presented in an Appendix to this chapter.

In addition to the correction of faulty posture and movement, the Mensendieck principles are also applicable to preparing for childbirth, and to postpartum and postsurgical and posttraumatic reconditioning. It also provides excellent preparatory and supplementary training for acting, dancing, modeling, yoga, and aerobics, as well as for all sports.

Since no vigorous exercising is necessarily involved, Mensendieck has proven to be of exceptional value for cardiac, neurologic, and orthopedic patients. It is also highly suitable for working with children during their formative years, and with elderly individuals.

Teaching is usually done on an individual basis. This enables the instructor to adapt his guidance to the physical and mental make-up of the individual. Occasionally, advanced pupils can work in small groups. It is not uncommon to find pupils taking lessons once a week for years and years. Although the posture and the muscle functioning of these pupils have already reached perfection, they enjoy the challenge of continually learning and executing the more complex movements which their Mensendieck specialist can provide.

In the United States too, the popularity of The Mensendieck System is steadily increasing, as its unique merits become more widely known. Many Americans hear about Mensendieck while in Europe; others have read Dr. Mensendieck's books.

Because of the over-all scarcity of Certified Mensendieck Specialists in the United States, our home country since 1959, this book is our attempt to comply with the many requests we have received from the general public over the years for more information pertaining to our imported profession, preferably in the form of a thorough self-instruction guide for beginners.

APPENDIX TO BACKGROUND INFORMATION

DETAILED EXAMINATION REQUIREMENTS FOR MENSENDIECK SPECIALISTS IN THE KINGDOM OF THE NETHERLANDS

All examinations for Mensendieck teachers take place under the supervision of government representatives from the Department of Social Affairs and Public Health.

The A Examination: The examination in *ANATOMY* concerns the structure of the human body in general and of the movement and posture mechanisms in particular. It includes the following subjects:

Cytology—cell and cell division.
Histology—epithelial tissue; connective tissue, including adipose tissue, cartilaginous tissue and bony tissue; muscles; nerves; embryological development as far as it applies to the further study of anatomy and physiology.
Microanatomy—principles of microscopic anatomy of the major body organs, such as glands, liver, pancreas, G-I tract, lungs, ovaries, testes, uterus, etc.
Descriptive anatomy—general osteology: the bones; arthrology: the joints, freedom and restriction of movement, joint mechanisms, structure of the spinal column, rib articulations, the extremities, etc.; myology: smooth and striated muscles, structures of muscles as related to function, muscle mechanics, etc.
Topographic anatomy—position, form, and relation to surrounding parts of the thoracic and abdominal organs.
Gross anatomy in vivo—the external form of the human body.
Analytical anatomy—analysis of posture and motion in their relationship to functional exercises.

*Examination Requirements in effect in 1972.

The B Examinations: The B examinations consist of a series of individual examinations in each of the following subjects:

I. *NEUROLOGY*—anatomy and physiology of the central and peripheral nervous systems; clinical neurological examination and testing methods; knowledge of those congenital and acquired conditions of these nervous systems of particular importance to the Mensendieck kinesio-therapist.
II. *PHYSIOLOGY*—general introduction, including physical and (bio)chemical principles; general physiology; metabolic processes; digestive processes; water balance and kidney function; respiration; functions of the integument; thermal regulation; functions of the blood, heart, and vascular system; endocrinological functions; muscular physiology; pregnancy and birth; physiology of vision and hearing.
III. *ORTHOPEDICS*—general principles of orthopedics; examination and testing techniques; principles of the treatment of congenital, developmental, and acquired deficiencies; inflammations; traumata; degenerative and neoplastic diseases of the body's support and movement structures; neurological diseases of particular importance to orthopedics; the orthopedic diseases, with emphasis on their static and static-dynamic aspects.
IV. *PATHOLOGY*—general pathological principles and definitions; pathologenetic factors and immunology; localized deficiencies of nutrition, metabolism, and functions; generalized and specific infections; regenerative processes and repair; general pathology of neoplasms and systemic diseases; general deficiencies of nutrition and metabolism; pathology of movement and locomotion related to internal medicine; pathological conditions of the heart, vascular system, lungs, gastrointestinal tract, liver, pancreas, kidneys; blood grouping and transfusions; endocrinological dysfunctions.
V. *THEORY OF THE MENSENDIECK SYSTEM*—this subject includes (1) knowledge of all fundamental postures, positions, and movements, in accordance with the theoretical

principles of the Mensendieck method; (2) the ability to systematically analyze all Techniques relative to the muscle functions and skeletal movements involved, and the application of the basic principles of the Mensendieck Techniques in everyday movements.

VI. *APPLICATION OF TEACHING PROCEDURES ACCORDING TO THE MENSENDIECK METHOD*—demonstrates the student's ability to practically apply the theoretical principles of this method.

VII. *PREGNANCY AND BIRTH*—requires in-depth knowledge of the processes of human pregnancy and normal childbirth, and the ability to instruct specific Mensendieck Techniques and apply certain Mensendieck principles prior to, during, and after childbirth.

VIII. *TREATMENT ACCORDING TO THE MENSENDIECK METHOD*—demonstrates the student's insight in the physical capabilities and limitations of an individual patient as this pertains to the patient's expected ability to respond favorably to treatment based on the application of the Mensendieck principles, and to ensure that these principles are applied correctly.

IX. *THE SELF-EXECUTION OF, AND INSTRUCTING OTHERS IN, THE MENSENDIECK TECHNIQUES*—during this examination the student has to demonstrate her knowledge, experience, and degree of perfection attained during her training in all fundamental posture and movement Techniques contained in the Mensendieck System, and must demonstrate her teaching approach and ability.

The A examination constitutes a three-hour written examination. The B examinations consist of a two-hour written examination for each of the applicable subjects. The oral examinations in each of these subjects take twenty minutes. The practical examinations (B VIII and B IX) require approximately an hour and a half for both of these subjects together.

In addition to the above requirements and before the Mensendieck student is admitted to participate in the B examinations, she must have first successfully passed the A examination, and she must have shown proof of having attended at least thirty classroom hours in general psychology, twenty hours in public health, and twenty-five hours in the principles of physical education, all at the academic level.

To follow a training course offered at the Mensendieck Institute in Amsterdam, 1) Command over the Dutch language is required as a necessity for a non-Dutch student to follow and give instructions, and 2) Permission to follow the course has to be requested from the Minister of Public Health and Environmental Hygiene, 8-12 Dr. Reyersstraat, Liedsendam, the Netherlands. The application must be accompanied by photocopies of the High School diploma and/or Bachelor's degree with majors in physics, chemistry and biology. Copies of the diplomas and judicia also have to be sent to the President of the Stichting tot Opleiding van Oefentherapeuten Mensendieck, 1-3 Nieuwe Vaart, Amsterdam, the Netherlands.